Garden Practices and Their Science

Written in a clear and accessible style, *Garden Practices and Their Science* guides gardeners in the practical arts of plant husbandry and in their understanding of its underpinning principles. The author, Professor Geoff Dixon, is an acknowledged and internationally respected horticulturist and microbiologist; he intertwines these arts and principles carefully, expertly leading readers from one to the other.

Achieving the manipulation of plant life is described in eight full-colour, well-illustrated chapters covering the growing of potatoes, bulb onions, legumes, small-seeded vegetables, soft fruit, bulbs and herbaceous ornamentals in great detail. Environmental factors controlling the successful husbandry of these crops is described in simple, non-technical language, increasing gardeners' enjoyment and competence. Gardeners are also informed of the tools and equipment they require and their safe use. Also provided are a series of simple, straightforward tests identifying the aerial and soil environments beneficial for plant growth using readily accessible domestic tools. Discussions of very straightforward techniques for vegetative propagation conclude this book. Each chapter ends with a list of the gardening knowledge that has been gained by readers. The structure of this book fulfils a longstanding need for descriptions of practical skills integrated with the corresponding biological reactions of plants. Emphasis is placed on gardeners' development of healthy soils, which encourage vigorous, active root systems capable of withstanding stresses—an aspect of gardening that rarely receives sufficient attention.

Tailored for readers requiring clear and concise directions, this very practical book is an instruction manual directed at early-stage gardening learners. These include people of all ages and requirements such as new garden owners, allotment-holders, apprentices and students of basic levels in the Royal Horticultural Society's or City & Guilds qualifications, career changers, community gardeners and those needing applied biological knowledge for GCSE examinations.

Geoff Dixon is Professor of Horticulture and an internationally distinguished biological and horticultural communicator, educator and research scientist. He has worked in Cambridge, then the Scottish Universities of Aberdeen and Strathclyde and Colleges of Agriculture, and is now a member of the School of Agriculture, Policy and Development (SAPD) at the University of Reading, UK.

Gardening's knowledge window

Garden Practices and Their Science

Geoff Dixon

Routledge
Taylor & Francis Group
LONDON AND NEW YORK

First published 2019
by Routledge
2 Park Square, Milton Park, Abingdon, Oxon OX14 4RN

and by Routledge
52 Vanderbilt Avenue, New York, NY 10017

Routledge is an imprint of the Taylor & Francis Group, an informa business

British Library Cataloguing-in-Publication Data
A catalogue record for this book is available from the British Library

Library of Congress Cataloging-in-Publication Data
Names: Dixon, Geoffrey R., author.
Title: Garden practices and their science / Geoff Dixon.
Description: New York, NY : Routledge, 2019. | Includes bibliographical references and index.
Identifiers: LCCN 2018031178 | ISBN 9781138485235 (hardback : alk. paper) | ISBN 9781138209060 (pbk. : alk. paper) | ISBN 9781315457819 (ebook)
Subjects: LCSH: Horticulture.
Classification: LCC SB318 .D58 2019 | DDC 635—dc23
LC record available at https://lccn.loc.gov/2018031178

ISBN: 978-1-138-48523-5 (hbk)
ISBN: 978-1-138-20906-0 (pbk)
ISBN: 978-1-315-45781-9 (ebk)

Typeset in Rockwell and FranklinGothic
by Apex CoVantage, LLC

To my wife, Kathy Dixon, and to Lucy, Richard, Dougal and Amanda, and my grandchildren Isabella and Hector McNeill and James Dixon.

Contents

Illustrations

Figures

Tables

Preface

Sharing the joys and stimulation of gardening is the aim of this book. Gardening gives great pleasure plus mental relaxation and physical stimulation. Working with and observing plant growth reveals the intricate biology happening before the gardener's eyes. Combining gardening practices with seeing and feeling why and how plants deliver pleasure is the substance of this book. Appreciating living relationships between gardening practices and their underlying horticultural science deepens the returns from an already fruitful occupation. The gardener then knows why his or her skills make plants perform, produce and fruit.

This book encourages the enthusiasm of people newly involved in gardening. These may be early-stage learners, people attracted into gardening by career changes or new hobby gardeners, intent on cultivating their own land or an allotment. Intertwining an understanding of practices and processes dispels the mystique sometimes associated with gardening.

Gardeners are a wonderfully wide and diverse fraternity who willingly embrace new entrants, for whom this book is written. Keen sharing of information, knowledge and plants typifies this fraternity of knowledgeable hobbyists, skilled professionals and expert horticulturists well versed in why and how plants are grown. New entrants should never hold back from asking seemingly naïve questions. Not infrequently, these identify problems which continue perplexing even the "experts". The adage "the only dumb question is the one you don't ask" is good guidance.

Perspectives of 70 years of gardening experience are invested in this book, along with deep desires for an understanding of why plants grow as they do. One of my great mentors in a voyage into professional horticulture was Mrs Audrey Dixon (no relation), a gifted biologist. She taught me how my interests in gardening could be translated into science, and particularly into botany, chemistry and zoology for London University's Advanced and Scholarship studies. My parents, Thomas and Nora Dixon, sacrificed much in encouraging that translation at Pewley Grammar School in Guildford. Studying at Wye College (London University's Faculty of Agriculture and Horticulture) deepened immeasurably my integration of biology and plant performance, stimulated by my mentor Professor Herbert Miles.

My dear wife, Kathy Dixon, has supported and sustained me in science, professional horticulture and gardening for well over 50 years. My good friends and colleagues, Dr Robert Bogers (formerly, Director of the Bulb Research Station, Lisse, the Netherlands), Michael Hall (formerly, Deputy-Principal, Writtle University College) and Dr Ian Tribe (formerly, Reader in Biology, University of Surrey) have improved this book's entire text immeasurably with their careful, critical comments and suggestions. Friends and colleagues in the School of Agriculture, Policy and Development of the University of Reading, especially Professors Paul Hadley and Helmut van Emden, and the late Professor Sir Colin Spedding, provided additional very welcome interest and support. Centres of my professional career in Cambridge at the National Institute of Agricultural Botany and in the Universities of Aberdeen and Strathclyde integrated with Scottish Schools and Colleges of Agriculture provided additional sources of knowledge and experience in biology, horticulture and gardening contributing to this book.

Sherborne, Dorset; June 2018

Acknowledgements

Frontispiece: gardening's knowledge window, designed and built by Sallis Chandler, Landscape Designers, for the Floral Market exhibited by Marks & Spencer plc which was awarded a Gold Medal at the Royal Horticultural Society's 2018 Chelsea Flower Show

P 1.1 Mr James Hadley cartoon

P 1.2 Wilkinson Sword well-balanced spade

1.8 Mr James Hadley, gathering and using energy

1.10 Transport services ex figure 6.17 "phloem loading and unloading" in Adams et al. *Principles of Horticulture Level 3*; published by Taylor & Francis.

2.42 Mr Dougal McNeill, soil erosion by water

2.58 Dr Glynn Percival, soil auger

3.1 Seed companies: Marshalls (includes Unwins), Mr Fothergills and D. T. Brown, Thompson & Morgan

3.2 Thompson & Morgan, vegetable seed packets

3.56 Thompson & Morgan, flower seed packets

3.56 Comparison of mitosis and meiosis; Figure 12.7 "comparison of meiosis and mitosis" Raven, P., & Johnson G. B. (1999). *Biology*, 5th edn, London: WCB/McGraw Hill

3.57 Generalised flower structure; Figure 8.3, from Adams et al., *Principles of Horticulture Level 2,* Taylor & Francis

3.66 Life cycle of plants, courtesy of Professor Paul H. Williams, Department of Plant Pathology, University of Wisconsin, Madison, USA

4.1 Table calendar of seed sowing and growing, adapted from vegetable seed growing guide, Marshalls Seeds Ltd, Cambridgeshire

4.48 Dibber Spear & Jackson, Sheffield, South Yorkshire

4.54 Basic plant structure; Figure 34.2 "Diagram of a plant body", Raven, P., and Johnson, G. B. (1999). *Biology*, 5th edn, London: WCB/McGraw Hill

4.55 Tree rings, courtesy Dr Glynn Percival

4.56 Tree rings close-up, courtesy Dr Glynn Percival

4.61 Sources and sinks ex Figure 35.10. Diagram of mass flow, Raven, P., and Johnson G. B. (1999). Biology, 5th edn, London: WCB/McGraw Hill

5.41 'Big Ben' blackcurrant ripe fruit, courtesy Dr Rex Brennan and the James Hutton Institute, Invergowrie, Dundee

5.44 Unnamed green blackcurrant fruits, courtesy Dr Rex Brennan and the James Hutton Institute, Invergowrie, Dundee

5. 51 Internal structure of monocotyledon and dicotyledon stems; Figure 6.4 from Adams et al., *Principles of Horticulture Level 2*, Taylor & Francis

5.52 External structure of monocotyledon and dicotyledon plants; Figure 7.2 from Adams et al., *Principles of Horticulture Level 2*, Taylor & Francis

6.28 Diagrammatic section of *Narcissus* bulb; Figure 7.21 structure of a *Narcissus* bulb from Adams et al., *Principles of Horticulture Level 2*, Taylor & Francis

7.34 Long short and day neutral flowering; Figure 39.23, how flowering responds to day length. Raven, P., and Johnson G. B. (1999), *Biology*, 5th edn, London: WCB/McGraw Hill

8.35 Mist bench atomiser, courtesy Wright Rain (MacPenny), Southampton, Hampshire

8.36 Mist sensor, courtesy Wright Rain (MacPenny), Southampton, Hampshire

8.37 Mist sensor and cable, courtesy Wright Rain (MacPenny), Southampton, Hampshire

8.39 Control box and plants, courtesy Wright Rain (MacPenny), Southampton, Hampshire

8.40 Weaning plants after propagation using mist, courtesy Date Palm Developments, Wells, Somerset

Mr Michael Burks, owner of the Gardens Group, Sherborne, for permitting photographs of displays to be taken in his garden centre, the Garden Centre Group, Sherborne, Dorset

Glossary

Scientific, horticultural and gardening terms used throughout this book that are not in common use are defined and explained here. Where a term or abbreviation occurs frequently, it is defined in this list; where it occurs only once, it is defined at its occurrence in the text.

adenosine triphosphate (ATP) the molecular unit of currency for energy in cells, it changes to adenosine diphosphate (ADP), following energy discharge.

analysis the reduction of larger compounds to smaller, simpler ones, with the release of energy which is used in work processes such as growth, flowering and fruiting.

apical dominance growth by the terminal bud of a shoot which suppresses the development of lateral buds.

asexual reproduction that does not involve sexually produced gametes. In plants, this usually means encouraging the development of roots on plant parts either before or after they are detached from the parent. Rooted detached parts then form independent plants.

bactericides substances that kill bacteria.

biomass the total matter composing an organism, most commonly used in the broader sense covering a whole population or the inhabitants of a region.

calyx the outer part of a flower, usually consisting of green (chlorophyll-containing) leaf-like structures, termed sepals, which in the bud stage enclose and protect the inner parts.

catalyst a substance that speeds up chemical reactions without itself undergoing permanent change.

chemical ion an electrically charged particle that has gained or lost electrons.

chemical symbols a shorthand code for chemical elements; the earliest known elements have Greek or Roman names and symbols, for example gold has the symbol Au from the Latin *Aurum*.

chloroplast an *organelle*, known as a plastid, in plant cells that contains the pigment chlorophyll; this is where photosynthesis takes place. The organelles contain sac-like membranes that hold the chlorophyll.

conductivity in relation to soil and compost, the property of conducting electricity.

cytoplasm the jelly-like contents of a cell, excluding the nucleus, in which the *organelles* are suspended.

degree days the accumulated heating or cooling sum above or below a baseline at which a plant either commences or stops growth.

deoxyribonucleic acid (DNA) the thread-like chemical in the plant cell nucleus in which is coded all the molecular instructions for plant germination, growth, fruiting, senescence and death.

diploid describes a cell or nucleus containing two complete sets of chromosomes, one derived from each parent; in meiosis, the number of chromosomes is halved, forming the haploid state.

electron a subatomic particle with a negative charge.

element a pure substance that cannot be broken down into simpler materials.

enzyme a protein catalyst found in cells that is responsible for increasing the rate and maintaining the specificity of intra- and extracellular reactions.

etiolation the exclusion of light from green plants which causes blanching, stem cell extension and reduction in leaf size.

exudate fluids secreted from the aerial parts and roots of plants that contain amino acids, sugars, growth-regulating chemicals, and are now being recognised as important means by which plants communicate with the surrounding biological environment.

fallow the Middle English term "falow" describes land that is left unseeded for a period in the crop rotation; the land may be ploughed and cultivated before being left without cropping or left with the stubble from the preceding crop remaining.

fertility the ability of a soil to sustain plant growth and reproduction, frequently used in relation to agricultural, horticultural or garden cropping, it consists of three components, physical, chemical and biological factors.

fungicides substances that kill fungi and related organisms.

gamete a reproductive unit containing chromosomes in the haploid state having undergone reduction division (meiosis); the diploid state is restored when male and female gametes unite.

guard cells crescent-shaped cells which in pairs surround the *stomata*, opening and closing to control the entry and exit of gases and water vapour in leaves; mostly found on the underside of foliage, but they are also present in stems and other green organs.

habit the characteristic physical shape into which plants develop.

habitat the place or environment in which plants live.

herbicides substances that kill weeds; also known as weed-killers.

insecticides substances that kill insects.

lignin a long-chained molecule found as spirals in the cell walls of *xylem* vessels, making them strong, rigid and impermeable to water.

meristem an area of active *asexual* cell reproduction, the products of which form the permanent structures of plants, principally found at the tips of shoots and roots, also in the vascular tissues between the *xylem* and *phloem*, in young leaves and as a response to wounding.

mitochondria *organelles* present in the *cytoplasm* of cells where aerobic respiration takes place, regarded as the powerhouses of cells producing their energy currency, *adenosine triphosphate* (ATP).

molecule a group of similar or dis-similar atoms bonded together forming the smallest unit of a chemical compound.

nectary a gland in flowers that attracts pollinating insects by releasing sticky fluids containing sugars, amino acids and nutrients.

organ a functioning and anatomically individual unit of plants with a distinct activity, or activities, such as roots, leaves, stems, buds, flowers or fruits.

organelle a functioning and distinct unit within a cell; for example, *chloroplasts* and *mitochondria*, growing from pre-existing organelles and transmitted in maternal *gametes*.

osmosis the diffusion of water across a semipermeable membrane from a region of lesser concentration of ions to one of higher concentration.

pelleted seed a seed that has a protective coating, with a layer containing nutrients, *fungicides*, *insecticides* and beneficial microbes, which disintegrates in soil or compost as germination proceeds. Pelleting produces a rounded seed which is placed more easily and accurately by hand and by machine.

petal a modified leaf that protects the reproductive organs of flowers, frequently highly coloured as an attraction for pollinating insects; collectively, petals are known as the corolla.

phloem living cells in the vascular system of plants which conduct the products of photosynthesis from green organs of production (sources) to places where they are used or stored (sinks).

pistil the female organs of flowers consisting of the stigma, style and ovary.

plant growth regulators (hormones) chemical messengers capable of moving throughout plants that coordinate the germination, growth and reproduction processes.

ribose nucleic acid (RNA) a linear, single-stranded molecule transcribed from *DNA* which transmits the molecular instructions as a reference book or template from the nucleus to *ribosomes*, RNA is also present in *chloroplasts* and *mitochondria*.

ribosome specialised *organelles* in the *cytoplasm* where the instructions coded in *RNA*, obtained from the *DNA* in the nucleus, are translated into protein constructions which are then utilised in the germination, growth, reproduction and senescence of plants.

rootstock the underground parts of plants from which aerial organs are produced, a term most frequently used in the propagation of top fruit where the propagator constructs plants consisting of a rootstock onto which is united specially selected portions (scions) which will produce the fruit.

sexual reproduction that involves the union of *gametes* from male and female parents in the process of fertilisation.

spikelet part of a compound flower stalk or spike in which the flowers are directly attached to the central stem, typically found in the grass family (*Gramineae*).

stamen male organs of flowers consisting of a thread-like filament surmounted by the anthers where pollen grains, which contain the male *gametes*, are produced.

stomata openings on the surface of leaves and other green organs through which gas (oxygen and carbon dioxide) and water vapour exchange takes place with the atmosphere; opening and closing is controlled by the *guard cells* surrounding the stomata.

stubble the basal parts of herbaceous plants, especially cereals left after harvesting.

synthesis building larger compounds from simpler ones which store energy for future use.

taped seeds placing seeds at equal spacings on sticky, biodegradable tape is a convenient method of determining sowing density; avoiding requirements for thinning in the seedling stage, tapes can be loaded with *fungicides*, *insecticides*, nutrients or beneficial microbes as a means of promoting germination and healthy growth.

tepal an outer whorl of flowers where there is no distinction between sepals and petals.

tiller a side shoot growing from the base of a stem, typically found in monocotyledonous plants.

tissue groups of cells with common origin, structure and function.

transfer cells specialised cells with high surface to volume ratios because of surface infoldings which permit the rapid transfer of materials, especially sugar solutions, over short distances, commonly found connecting the vascular tissues of small veins in developing leaves and cotyledons.

translocation long-distance transfer of materials such as water, nutrients and the products of photosynthesis from entry into the plant or from the source of production to sinks where they are utilised.

transpiration the process by which water absorbed through the roots is actively passed out of *stomata* as water vapour, differing from evaporation since the rate is influenced by internal and external factors.

vascular tissue longitudinal strands of tissue composed of *xylem* and *phloem* tissues.

weight (mass) mass refers to the amount of matter in an object, while weight refers to the effects of gravity on an object; weight is used more frequently in common use.

xylem the part of the *vascular tissue* that conducts water and nutrients from plant roots upwards into stems, leaves, flowers and fruit. It also provides mechanical support against physical stresses and strains.

Consultation sources: Thain, M., and Hickman, M. (2004), *The Penguin Dictionary of Biology*. London: Penguin Books. Cross-checked with definitions supplied by Google online (2018).

Geoff Dixon

Preamble
Safety in the garden

Introduction

Like most other forms of physical activity, gardening involves using tools, machines and chemicals. These are perfectly safe if handled respectfully, but it is essential that a few very simple rules are observed for your protection and for that of others, especially children and pets.

a. Hand-tools

1. Treat all tools with respect and care, keep them clean and wipe over regularly with an oily cloth.
2. Small hand-tools should be carried in a box or plastic pot with the blades facing downwards.
3. Work with tools in front of you and with blades facing downwards in a comfortable and relaxed posture.
4. Small hand-tools should be worked by either bending over or by using a kneeler with the tool's blade facing downwards.
5. Move larger tools either in a barrow with blades facing down or carry them at the slope with the blades behind you.
6. Tools not in use should be placed back in the barrow or stood upright against a tree or wall with blades facing inwards; NEVER lay tools on the ground, especially with the blades or tines facing upwards (see Figure P1.1).
7. When lifting with tools, such as digging, lift by bending the knees not your back.

Figure P1.1 *Health and safety: the dangers of upward-facing tines*

b. Machinery

1. Garden machinery such as mechanised or electric lawn-mowers should be treated with special care; always have them serviced professionally at regular intervals.
2. Never place hands or feet near working parts of a machine, and always turn off the machine before making any adjustments or even removing grass cuttings.
3. Electrical cables should trail behind the machine's direction of travel; NEVER in front, and ensure there is a circuit-breaker installed between the machine and main power supply.
4. Fuel for petrol-driven machinery should be stored carefully in a garage or outbuilding in suitable metal or plastic cans in volumes not exceeding 10 litres. Decant into the machine using a metal funnel, avoiding spillages. Open the fuel tank on the machine first, decant, and then seal the petrol can before the machine's tank. This practice ensures that the greatest bulk of inflammable liquid is sealed as a priority. Never smoke or use naked flames in the vicinity of either the petrol can or the machine.

c. Children

1. Only children of sufficient size and maturity should be allowed to carry or use tools.
2. Children should only use garden machinery when strictly supervised and following detailed instruction and cautioning.
3. Children should not ride on garden machinery.
4. Carefully ensure that younger children are not tempted to eat unidentified fruits or other plant parts or place objects in their mouths.

d. Garden chemicals

1. As a matter of priority, wash your hands and face thoroughly after using garden chemicals.
2. Consider all the options for pest, disease and weed control before using chemical controls.
3. Chemicals sold for use in gardens have been very rigorously tested, ensuring their safety; the concentrations of active ingredients contained in garden chemicals are well below those used commercially; nonetheless, they should be used with care and caution.
4. Read the manufacturer's instructions before opening the container or packet and follow them rigorously when using the product.
5. Many chemicals come in sealed containers for direct use, do not tamper with this sealing.
6. Ensure that the recommended protective clothing is worn, this should include Wellington boots or very stout shoes, gloves and an eye-shield, and where advised also include water-proof leggings, coat and hat.
7. Children should not be allowed to use garden chemicals without thorough guidance and supervision.

8. Pets should be kept well clear of areas where chemicals are in use and where they have been used.
9. Do not exceed the manufacturer's recommended dose rates.
10. When spraying, ensure that the nozzle is directed downwards onto the areas requiring treatment; do not spray upwards since this may generate drift onto areas or plants not requiring treatment, and avoid spraying on windy days.
11. Spraying should be avoided on days with wind speeds greater than Beaufort scale 2 (light breeze at 3 metres per second).
12. All unused chemicals should be disposed of safely and not returned to the packet or other container.
13. Ensure that all sprayers are thoroughly cleaned and well washed with soapy water following each operation; wash out once with soapy water and then flush out twice with clean water.
14. Ideally, there should be separate and well-labelled sprayers for each type of chemical used, that is: weed-killers, fungicides and insecticides.
15. Keep a record of the garden chemicals used.
16. Contact a medical professional immediately if anyone using garden chemicals or who has been in their vicinity feels unwell, ensuring that a copy of the manufacturer's guidance sheet is available, as this will contain details of the active ingredient(s) and their antidote(s).

e. Glass

1. Gardening can involve the use of glass for greenhouses and cloches. This must be handled with care and preferably using suitably thickened gloves. Children should be kept well clear of glass structures at least until they are able to understand the risks involved and how these are avoided.
2. Glass breakages are inevitable and all broken glass should be collected immediately and disposed of safely.
3. Increasingly, manufacturers are using forms of safety glass which are preferable for gardening purposes.

f. Stakes and canes

1. Plants will require supports as they grow, possibly by using bamboo canes. The tops of canes should be guarded with plastic covers and stakes capped off with a suitably shaped piece of wood or by placing an inverted plant pot on top of the cane; this avoids inadvertent wounding, and especially damage to eyes when bending down in their vicinity.

g. First-aid

1. For all gardening operations, keep a well-stocked medical chest close to hand; kits sold by the Red Cross provide the basic essentials.

Figure P1.2 *A well-balanced spade*

Tools

Good-quality garden tools are a joy when used; poor-quality tools are sources of annoyance and tiring when working.

- Tools such as spades and forks should feel comfortable when picked up (see Figure P1.2, testing a good-quality spade for balance). The blades and tines should balance easily when the tool is held lightly three-quarters of the way down the shaft where the wooden handle and the metal collar are fitted together.
- Individual gardeners should decide whether they feel more comfortable with "D"- or "T"-shaped handles for spades and forks.

Essential tool kit

Digging spade, digging fork, bow rake, draw hoe, spring tine rake, garden line, hand trowel, gardeners' knife and sharpening stone, secateurs, hand shears, garden barrow or truck, watering can, stout gardening gloves, 4- or 5-ply garden twine.

Additional tools

Edging shears, edging knife, tree loppers, bow saw (used for pruning operations) onion hoe, four-pronged hand fork, two-pronged hand fork, daisy grubber, border fork, (additional tools for weed removal and cultivation), bulb planting trowel (allows planting at the correct depth), manure fork (eases handling).

Ancillary tools

Gardening involves various carpentry and metal working skills from time to time for which tools are needed. These include: Engineer's ball and pein hammer, claw hammer, club hammer, pliers, wire cutters, rechargeable electric drill (or drill with extension cables and circuit breaker), drill bit key, drills for wood and masonry, screwdriver, pick axe, felling axe, mallet, tape measure, square, supplies of suitably sized screws and nails.

Greenhouses, frames, cloches and plant supports

Seedlings may be started initially on window ledges, but unless the plants are very carefully monitored growth becomes distorted and spindly very quickly because of the low light levels and because it is coming principally from one direction. Indoor propagators fitted with fluorescent tubes or light-emitting diodes (LED) are available and will produce more robust growth.

The gardener will greatly increase his or her enjoyment, however, by purchasing some form of outdoor protected environment for germinating seed, starting seedlings into growth and preventing frost damage to maturing plants or those awaiting transplanting. The scale of provision is obviously related to the gardener's available funding. Ideally, a standard 2 × 3 m greenhouse fitted with benching and a roof ventilator would be adequate in terms of initial equipment and opportunities. Larger greenhouses benefit from being fitted with a side ventilator as an additional means of dispelling high humidity or excessive heat.

Alternatives such as free-standing polythene-covered cages or standard cold frames will offer capabilities for the early years of gardening and can be expanded as horizons and enthusiasm widens. Numerous gardening retail outlets sell these facilities at reasonable prices such as garden centres, the larger multiple retailers and web-based marketing companies.

If the gardener decides on purchasing a greenhouse, it should be orientated on a north to south axis. This allows the sun to pass evenly from east to west across the ridge of the roof, warming and illuminating the benches equally. All propagation facilities and greenhouses should be placed on a paved surface, as this allows easier access and working during wet and cold periods. Areas underneath the benches should be covered with pea gravel to depths of 10 to 15 cm as a means of providing adequate drainage for water coming from the benches above.

Once erected, the greenhouse or other facilities should offer good service for years to come. It is sensible to purchase structures that are more than adequate for the activities which are currently envisaged, because it will soon become evident that there are many additional uses for them and this usefulness expands over coming years. In periods of very hot weather, it is advisable to paint the outside of the glass with a proprietary "whitening". This reduces the internal temperature and helps avoid scorching damage for leaves and flowers.

Cloches are used for protecting plants growing *in situ* in the garden. Plastic netting and polyethylene cloches are widely available from garden centres and by mail order. The alternative for more experienced users are glass "barn"-shaped cloches which provide well-ventilated and protected environments, but should obviously be used with caution where young children or the very elderly are likely to be in their vicinity.

The gardener will also require sheets of plastic netting for protecting ripening fruit from bird damage and as supports for some climbing vegetables; the availability of rolls of wire netting may also be useful in this respect. Additionally, supplies of bamboo canes of various lengths will be needed as plant supports.

Watering cans

Watering cans can be purchased from garden centres or from web-based suppliers. The removable fitment at the end of the spout is termed a "rose". These are characterised as fine or coarse depending on the number and diameter of holes drilled into their surfaces; this regulates the size of water droplets emitted. Watering cans allow the water to be placed on or near garden plants at a carefully regulated rate which does not cause damage or erode the soil around them. Fine roses are used when delicate seedlings are being watered and the coarse rose is for more generalised purposes such as settling leek or strawberry plants into the ground.

Fertilisers

Gardening involves using a range of fertilisers, the most regularly mentioned one in this book is a general-purpose type frequently sold in garden centres and by mail order as "GrowMore" (see explanation below). The chemical composition of fertilisers and the meaning of labelling is described in Chapter 2. Frequently, fertilisers can be purchased most economically from agricultural merchants in bulk as standard field crop fertilisers packed in 25 kg bags. Alternatively, gardening and allotment clubs buy in bulk and distribute in smaller amounts to their members. Some gardeners may choose non-synthetic fertilisers, or materials termed "organic". This is a matter of personal preference.

The term "GrowMore" is adopted in this book not as an advertisement for any current or past commercial products. In the context of this book, it is a neutral term derived from the World War II *Dig for Victory* campaign, for which the then Government made available for gardeners and allotment holders "National GrowMore Fertiliser" with a standardised composition of 7 per cent of each of the major nutrients (as described in Underpinning knowledge 2.2, 4.4 and 7.3). That gave a 1:1:1 formulation (N:P:K), which was advised for use at approximately the equivalent of 30 g per square metre and provided balanced nutrition for most crops. Some crops, especially soft fruit, may require variations on this formulation, and this is indicated in the text.

Chapter 1
What makes plants tick?

Introduction

The Green Kingdom from which gardeners' plants come supports all life on Earth. Without the activities of green plants, and some bacteria and fungi, animal life, including that of human beings, would not be possible. Gardeners for about the last 10,000 years have increasingly gained and applied their knowledge, expertise and skills in controlling plant growth for food production, environmental enhancement and for people's physical and psychological comfort. It is these latter virtues of giving physical and psychological relaxation and enjoyment which initially attracts many newly interested gardeners and from where whole new careers may develop, or they simply fulfil the desires for enjoying a newly acquired garden.

Considerable self-satisfaction comes from germinating seeds, growing plants, encouraging their fruiting, and in the consumption of food that is owner-produced. Complete ownership of even a small part of your food chain from seed to mouth is an immensely satisfying challenge and achievement.

This chapter offers newly interested gardeners, early career entrants and others an understanding of some of the factors which make plants grow and mature. It provides a background explaining the basis of many of the gardening practices described in the succeeding chapters. Gardening is about controlling plant growth, and in achieving that the gardener provides the physical and chemical necessities for growth and reproduction. This chapter provides an understanding of what these necessities are, why they are essential and how plants make use of them. These necessities are some basic environmental constituents: water (moisture), air (oxygen and carbon dioxide), warmth, light, nutrients and a suitable substrate, usually soil or compost, for physical support. This introductory chapter examines these environmental factors and shows how a shortage or absence of one may limit plant development or stop it completely. The concept of "limiting factors is a fundamental principle that guides gardening success. Frederick Blackman, a British plant physiologist, in 1905 identified the *principle of limiting factors* as: "When a process is conditioned as to its rapidity by a number of factors, the rate of the process is limited by the rate of the 'slowest' factor.'' Understanding this without a requirement for working in the garden avoids some of the pitfalls and disappointments which newly interested gardeners often experience.

Garden practices

a. Basic ingredients for plant growth

Six essential factors are needed for successful plant growth, these are:

1. water (moisture),
2. air (oxygen and carbon dioxide),
3. warmth,
4. light,
5. nutrients,
6. substrates.

Water

Green plants each contain millions of compartments, or cells, all collectively contributing towards the overall growth and development of the entire organism. In each cell water is necessary for all the biological processes. Without water cells dry out and eventually cease functioning. Water is also the medium in which nutrients are absorbed from soil or other substrates on which plants are growing and moved around inside them.

Air: carbon dioxide and oxygen

Two gases in the planet's atmosphere are directly vital for plant growth, oxygen and carbon dioxide. Plants differ significantly from animals by having the capability for using carbon dioxide and turning it into more complex materials by one of Nature's most fundamental processes, photosynthesis (light-dependent synthesis) using energy taken from sunlight. In this natural construction process plants take carbon dioxide and build substances such as carbohydrates, fats and proteins directly for their own use and indirectly for that of others. This is why plants are crucially important as food sources for animals whether they are consumed first hand by herbivores or at second hand by carnivores.

In turn oxygen is required by both plants and animals. It is used in the cells by another of Nature's fundamental processes, respiration, which breaks down and reconstructs the materials formed during photosynthesis releasing energy and as a result growth and reproduction take place. All the chemical reactions either constructing or breaking down substances into simpler ones require sources of energy for all these processes of work taking place in cells. Photosynthesis harvests energy from sunlight and respiration releases it for further use. Both photosynthesis and respiration are two of the most ancient attributes of living organisms having been developed by the earliest organisms inhabiting the Earth.

Warmth

Temperatures between 5 °C and 35 °C are normally those at which plants grow and reproduce most effectively. Below 5 °C, the rate of these activities slows down very substantially, growth is impaired and plants eventually suffer from low-temperature damage. Similarly, above about 35 °C plants are damaged by excessive heat. The most efficient temperatures required by the processes of work within individual plants reflect the natural environments where they originally evolved. Broccoli for example, grows well at 15 °C, whereas sweet corn requires temperatures greater than 20 °C for satisfactory fruiting.

Light

In the natural processes of building simple materials into more complex forms green plants capture the energy from sunlight. This drives photosynthesis turning carbon dioxide into energy-rich compounds. Consequently, green plants are at the base of food chains providing herbivores and subsequently carnivores with their meals. Breaking down the complex compounds releases energy which the herbivores and carnivores use in doing work.

Nutrients

About twenty simple inorganic materials (elements), known as nutrients, are required by plants in their cellular building processes. Some nutrients are required in substantial amounts such as nitrogen (N), phosphorus (P), potassium (K) and calcium (Ca). Lesser quantities of other nutrients, known as trace elements, such boron (B), copper (Cu), iron (Fe), magnesium (Mg), manganese (Mn), molybdenum (Mo); sulphur (S) and zinc (Zn), are also needed, some in minute quantities, and in specialised circumstances. Atmospheric nitrogen cannot be used directly by plants and must be converted into compounds with oxygen or hydrogen before it is taken up by roots. The individual elements are incorporated into cellular products or used in driving particular facets of growth, development and reproduction. Nutrients are taken into the green plant via the roots from the soil or an artificial substrate as discussed in Chapter 2.

Substrates

Plants require forms of stable physical support into which their roots can grow, establishing anchorage and from which they obtain supplies of water, oxygen and nutrients. The commonest substrate is soil which contains mineral nutrients. Additionally, in the soil roots encounter a vast range of microbes and other larger organisms. Some of these are highly beneficial and while others, fortunately far fewer, are extremely damaging causing plant diseases or feeding off plant parts as pests. Artificial substrates such as composts may contain microbes, or have them added, which support green plant growth. Alternatively, some substrates used by gardeners are inert, such as the expanded clays, perlite and vermiculite. Both biologically active and inert substrates must offer opportunities for root penetration by which the plant is supported. In some circumstances, gardeners provide additional physical support in the form of canes or posts and wirework (see Chapter 5). Stout supports are necessary for example, when gardeners grow plants in liquid cultures, known as hydroponics. Water culture or hydroponics was introduced as a term by William Frederick Geriche working at the University of California, Berkeley, USA in 1937.

b. Making plants grow

The effect of environment factors on plant growth can be seen in some simple window-ledge gardening.

b1. Showing the effects of water, light and warmth

Packets of broad bean seed can be obtained from a local garden centre or from a mail-order seed supplier. Broad beans are some of the larger seeds and hence are easily handled and used in this demonstration. These seeds are capable of being stored until needed if placed in a tin or similar container and placed in a cool, dry, well-aired cupboard. They should be kept well away from sources of heat such as radiators.

The kit needed includes:

1. 6 clear plastic drinking cups,
2. sheets of plain white kitchen roll,
3. tweezers,
4. water,
5. kettle,
6. large kitchen bowl,
7. scissors.

Two cups are required for each of the following environments:

1. warm, well-lit window sill (or a greenhouse bench),
2. unlit but warm cupboard such as an airing-cupboard,
3. shelf in a domestic refrigerator.

Beforehand the kitchen roll should be cut or folded such that it fits snugly into the drinking cups as double-folded pockets and stands against the walls. Half the cups should have water added into the bottom 1 cm and allowed to seep up into the paper refilling as necessary. The other cups are kept dry.

1. Count out sufficient broad bean seeds so that three can be placed in each cup.
2. Soak half of these seeds in tepid water, not hotter than 15 °C in the bowl.

The seeds will absorb water, swell and soften over a period of 30 to 60 minutes. Remove any seeds that rise to the surface, as these will not be viable and should be rejected. This demonstrates the importance of water. Without it, the seeds will not commence growth.

Once the seeds have swollen, they may be inserted gently using the tweezers in between the folded blotting paper in each cup.

Also set up the cups where the blotting paper or kitchen roll has not been wetted with dry seeds inserted into the folds; hence these are dry seeds in a dry environment.

Place one each of the "wet" and "dry" cups into the selected environments.

Inspect each pair of cups on a daily basis. Within 5 to 7 days results should begin showing.

1. Seeds placed in the dry cups which received no water will remain inert. In the absence of water, the seeds cannot commence development despite having adequate warmth and light.
2. Similarly, dry seeds in the cups placed on the refrigerator shelf will not develop. When the seeds have been imbibed and have access to water, growth will be exceedingly slow. Most refrigerators operate at about +4 °C, which should either prevent or at least slow growth significantly.
3. Seed in those cups placed on a well-lit, warm window sill (or a greenhouse bench) which have received water will sprout and produce a developing root system and extending shoot which should grow beyond the top of the cup.
4. The dry beans placed in the darkened cupboard (or airing cupboard) should not show any signs of development.
5. In the wet cups the seeds placed in the darkened cupboard (or airing-cupboard) will sprout producing roots and possibly a shoot. But this will be severely bleached and yellowed demonstrating that in the absence of light green plant growth is stressed.

See Figure 1.1.

Figure 1.1 *Broad bean seedlings with and without light and warmth; the left-hand bean has grown in warmth, moisture and light; central bean has grown in warmth and water but without light; and the right-hand bean has not received warmth, moisture or light, and hence shows no signs of growth.*

This demonstrates the need for water, warmth and light for the development of plants from seed. Broad beans are useful because they are large, easily handled, germinate relatively quickly and are easily observed.

c. The need for nutrients and physical support

The test plant is cress or mustard, seed of which can be obtained from the garden centre or mail-order sources.

Kit needed includes:

1. small plastic food trays with holes in the bottom,
2. a bag of seed-sowing compost obtained from a garden centre,
3. some sheets of kitchen roll,
4. water,
5. liquid plant feed diluted tenfold in a kitchen measuring jug,
6. small garden sieve,
7. large tray capable of holding all the smaller trays which does not have holes in the bottom,
8. small watering can with a fine rose attached,
9. a warm, well-lit window sill, or a greenhouse bench.

c1. Demonstrating healthy growth

Procedure

1. Fill two food trays with compost within 1.5 cm depth of compost. Tapping the container on a firm surface will settle the compost.
2. Sprinkle a small quantity of cress or mustard seed evenly over the surface of each tray, ensuring that they have sufficient space around each for individual development.
3. Through the garden sieve, sprinkle an even layer of compost to 0.5 cm deep over the surface of the trays. Careful sieving will ensure that the correct depth of compost is achieved which covers the seed.
4. Water the trays gently using the watering can, ensuring that this evenly moistens all the compost. Turn the rose upwards so that water is applied gently.
5. Place the trays in the larger tray and stand the whole on a window sill (or on a greenhouse bench).

Within 3 or 4 days the seeds will begin germinating and should then grow steadily producing healthy leaves. stems and roots (see Figures 1.2 and 1.3).

c2. Demonstrating the need for nutrients

1. Repeat the process of filling two trays with compost as described in c1.
2. Place folded sheets of kitchen roll in the bottoms of two more trays.

Figure 1.2 *Healthy cress seedlings*

Figure 1.3 *Healthy cress seedlings supported on compost. The side of the tray has been cut away, revealing the healthy white root systems that have penetrated throughout the compost, which provides the plants with nutrients and support.*

3. Sprinkle cress seeds evenly across the surface of the compost and kitchen roll paper.
4. Moisten the compost and kitchen roll paper gently with water from the watering can.
5. Stand the small trays in another larger tray.
6. Place all the trays on a well-lit window sill (or greenhouse bench).

This exercise will differentiate between growth with and without a nutrient source, the compost.

Monitor the progress daily. Should the surfaces of the compost or kitchen roll paper start drying out then re-wet them using the watering can. The kitchen roll paper will dry out more quickly than the compost.

1. Growth should be evident with the trays filled with kitchen roll paper within 3 to 4 days. This will be seen as the seeds beginning to swell followed by the development and growth of roots and shoots.
2. The trays containing compost should begin showing seedling shoots emerging through the surface within 10 days. These should be very robust and healthy plants because they are drawing nutrition and support from the compost.
3. The seedlings growing on kitchen roll will by this time show signs of stress because they lack a source of nutrition and since they are poorly supported as compared with those where roots are growing into the compost.

This compares seedlings grown on compost with those grown on kitchen roll paper. Compost provides the seedlings with nutrients whereas the inert kitchen roll does not (see Figure 1.4).

Figure 1.4 *Cress seedlings with and without nutrition, grown on inert kitchen roll (left) and seedling compost (right)*

c3. Demonstrating the effects of adding nutrients to seedlings grown on kitchen roll

1. Set up four trays lined with kitchen roll and sown with cress or mustard seed.
2. Take 10 ml of proprietary nutrient solution and dilute to 100 ml using a kitchen measuring jug.
3. Moisten the kitchen roll with plain water for two trays and with very dilute nutrient solution for the other two.
4. Place on the window sill (or a greenhouse bench).

Within 2 to 3 days germination and growth will commence. Seedlings receiving the diluted nutrient will grow more rapidly and healthily compared with those receiving plain tap water. Some growth does, however, continue with the latter seedlings because tap water contains some nutrients (see Figure 1.5).

d. Demonstrating the need for carbon dioxide in plant growth (photosynthesis)

The kit required includes:

1. two well-washed and dried plastic drinks bottles made of clear plastic, one with a tightly fitting screw top,

Figure 1.5 *Healthy cress seedlings growing robustly having received nutrient (left-hand side), and the much-reduced growth of those provided with tap water (right-hand side)*

2. seedling-sowing compost,
3. cress or mustard seed,
4. water,
5. soda-lime or activated charcoal (soda-lime is available as Spherasorb from Intersurgical contacted via the Amazon website),
6. small nylon netting bags, cut and formed from discarded women's tights,
7. cotton sewing thread.

Note: Activated charcoal absorbs carbon dioxide but more slowly than soda-lime. Activated charcoal is available from pet shops and garden centres dealing in sundries for aquaria or by mail order. Soda-lime is a granular mixture of calcium hydroxide with sodium hydroxide or potassium hydroxide which absorbs carbon dioxide more quickly. Surgical gloves should be worn when handling both materials.

1. Place a small quantity of un-moistened compost in the bottom of each plastic bottle and sprinkle in some seed on top.
2. Moisten the compost by adding a small quantity of water.
3. Leave the bottles unsealed and placed on the well-lit window sill (or a greenhouse bench).
4. Monitor on a daily basis adding small amounts of water if necessary (beware over watering as the seedlings will become infected by fungi and rot).

5. When the seedlings have germinated and emerged and are growing strongly, suspend a bag of soda-lime inside one bottle suspended with cotton thread and seal the top tightly with the screw cap.
6. Leave the other bottle unsealed and without soda-lime.

The soda-lime will over the following 8 to 10 days remove all the carbon dioxide from the atmosphere inside the bottle. The seedlings in this bottle cannot therefore, carry out photosynthesis and become yellowed and begin shrivelling. By comparison seedlings in the other bottle will continue growing and expanding normally (see Figures 1.6 and 1.7).

e. Demonstrating the carbon dioxide produced by respiration

Demonstrating the requirement for oxygen in respiration requires laboratory apparatus and several chemicals. The respiratory process can be demonstrated, however, by showing that it releases carbon dioxide. That is the converse of photosynthesis which absorbs carbon dioxide. The simplest way of achieving this demonstration is by showing that the microbes present in soil respire, producing carbon dioxide, in a similar way to green plants.

Figure 1.6 *Cress seedlings grown with carbon dioxide (left-hand side) and without carbon dioxide in a bottle (right-hand side)*

Figure 1.7 *Cress seedlings close-up showing seedlings with carbon dioxide (left-hand side) and stress from being deprived of carbon dioxide (right-hand side)*

Kit required for this demonstration:

1. two samples of garden soil, one sample should be kept in its natural state, the other should be exposed to high temperatures (greater than 100 °C),
2. place the soil to be heated on a tinfoil tray and stand in a domestic oven for 30 minutes,
3. two well-washed and dried plastic drinks bottles made of clear plastic with well-fitting screw tops,
4. solution of calcium hydroxide (lime water),
5. small nylon netting bags, cut and formed from discarded women's tights,
6. cotton sewing thread.

Note: Lime water is prepared by placing 15 g of calcium hydroxide in 300 ml of water in a basin, stirring vigorously with a mechanical stirring device (for example, an "Aerolatte® mooo"), and leaving for 24 hours. Decant the clear lime water into a plastic beaker or drinking cup.

Calcium hydroxide is known as hydrated or slaked lime and is available in some garden centres or by mail order. It is a very finely powdered type of liming material and hence more rapidly acting compared with more traditional ground calcium carbonate (chalk) materials. (Pharmaceutical-grade calcium hydroxide is available in 100 g quantities from Source Chemicals contacted via the Amazon website.)

Place about 75 ml of lime water in each of the plastics drinks bottles. Place samples of the living and killed soils in bags made from the women's tights and suspend inside the drinks bottles by cotton threads. Seal the bottles tightly with their screw lids.

After 7 to 10 days, the lime water in the bottle containing living soil will become clouded because the carbon dioxide coming from the respiring microbes has reacted with the calcium hydroxide, producing calcium carbonate (chalk).

No carbon dioxide is produced by the dead soil hence the lime water will remain clear, although some slight deposit of chalk may appear because of the carbon dioxide present in the atmosphere of the bottle. For a comparison between respiring microbes in living soil with cloudy lime water and the clear lime water in the bottle with dead microbes, see Figure 1.8.

Figure 1.8 *Comparison of killed microbes and the absence of carbon dioxide (left-hand side) and living soil-borne microbes respiring and producing carbon dioxide (right-hand side) where the calcium hydroxide and carbon dioxide have reacted, producing calcium carbonate (chalk)*

This chemical reaction may be described by the following:

calcium hydroxide + carbon dioxide →→→ calcium carbonate (chalk) + water

All these demonstrations give gardeners an appreciation of the four basic biological processes which are fundamental for successful gardening. These are: photosynthesis, respiration, transportation and growth, processes which are now described in more detail.

Underpinning knowledge

1.1 Gathering energy from sunlight (photosynthesis)

Photosynthesis is the natural process by which plants harvest energy from light, principally sunlight. Energy is used for building, or synthesising, atmospheric carbon dioxide into sugars. This process releases oxygen. Gardening which encourages the cultivation of green plants is one of the principle means by which the energy in sunlight is harvested for the benefit of mankind.

Energy is measured in calories. This is one of the units used by utility companies when describing the heating capabilities of domestic gas. One calorie of energy will raise the temperature of one gram of water by one degree Celsius (1 °C). Energy is used by plants for doing physical, chemical and biological work. Sunlight provides the earth with 13×10^{23} calories per year, or 40 million billion calories per second. Plants harvest this energy by photosynthesis, but even so they capture only about 5 per cent of the Sun's energy which reaches Earth.

The term "photosynthesis" was coined by the American botanist Charles Reid Barnes in 1893 describing the building by green plants of complex carbon compounds from carbon dioxide and water in the presence of a green pigment, chlorophyll, and energised by sunlight. The processes can be summarised very simply:

Energy from sunlight

Carbon dioxide + water →→→ simple sugars + oxygen

A comparison of the processes of photosynthesis (energy capture) and respiration (energy use) is shown in Figure 1.9.

Water is taken into plant roots and drawn up through into the stems and leaves. Carbon dioxide is drawn into the leaves through pores known as stomata. These are found largely on the under surfaces of leaves and may also occur in stems and some parts of flowers. They open during daytime and close at night. Rates of photosynthesis are related to the concentrations of carbon dioxide in the atmosphere, temperature, the intensity of sunlight as an energy source and the availability of water as limiting factors. Rates of photosynthesis decline as leaves age, when they begin losing their deep green colour, turning yellow and senescing.

The energy contained in sunlight is captured during photosynthesis by the green pigment, chlorophyll, which is present in the cells of leaves and young stems in structures, organelles,

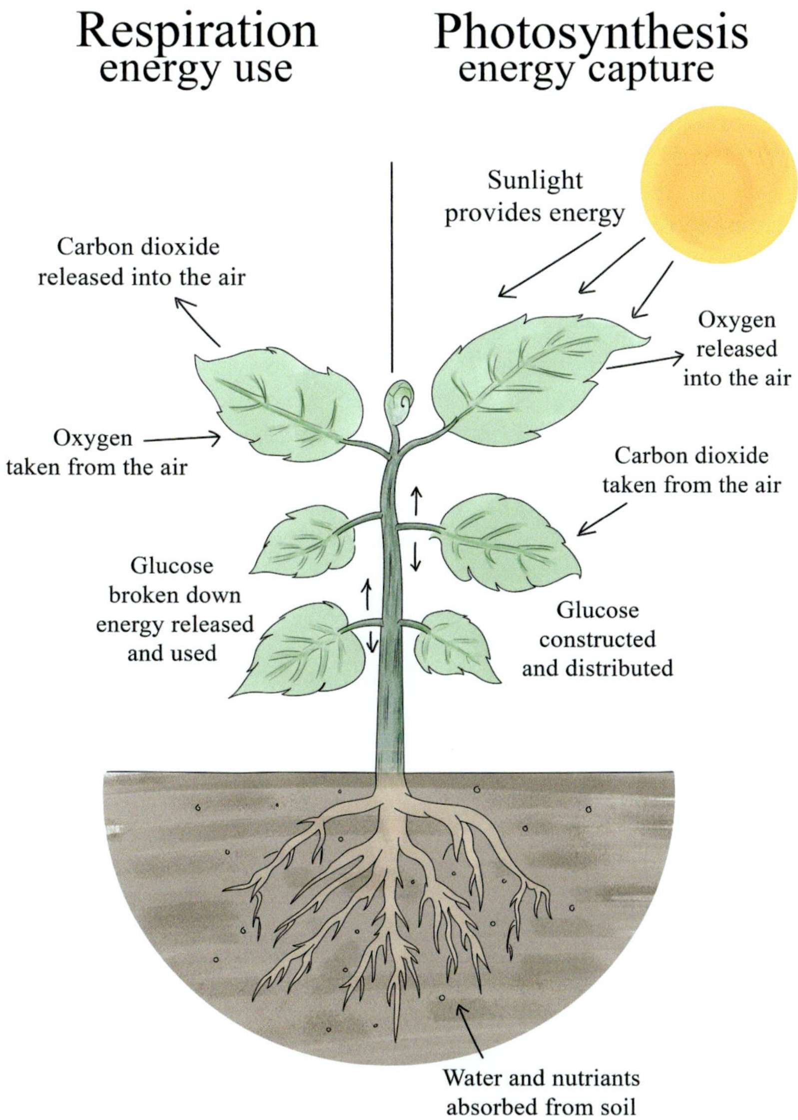

Figure 1.9 *Comparison of energy-gathering from light (photosynthesis) and using energy (respiration)*

called chloroplasts. There are between 10 and 100 chloroplasts in each plant cell, the number depending on its position within a leaf, stem or flower and their ages.

Carbon fixation during photosynthesis takes place in three stages:

1. energy capture,
2. moving energy,
3. making chains of carbon atoms of increasing length and complexity.

The sunlight arriving into a leaf is a series of energy packets. Those packets coming from the shorter wavelengths of light have the greatest energy levels. Energy is taken from the packets in the chloroplasts and passed onto a carrier molecule known as adenosine triphosphate (ATP). This molecule has abilities for moving energy into a series of steps in which carbon atoms are linked together in chains of increasing length forming initially simple sugars such as fructose and glucose. This leads into the formation of lengthening chains of carbon atoms from which more complex sugars, amino acids and fatty acids are manufactured. These are used further in plants as components for building carbohydrates (starches), proteins and lipids (fats).

Chlorophyll contains atoms of element magnesium, the energy carrier molecules (ATP) contain phosphorus and amino acids contain nitrogen and in some cases sulphur. These are essential elements for the functioning of photosynthesis and enter plants as nutrients from the soil via the roots. They are subsequently transported upwards into the leaf cells where photosynthesis takes place. This is one reason why the adequate supply of nutrients is essential for plant growth.

Carbon fixation in photosynthesis like all chemical processes in plants is dependent on the presence of enzymes. These are specialist proteins which drive reactions without being chemically involved in them; they are naturally occurring catalysts. Some enzyme-driven processes also require the assistance of other supporting molecules. This is the case with energy transfers involving ATP where an assister molecule, or co-enzyme, is involved. Constructing the chains of carbon atoms also involves a third and particularly important enzyme, ribulose biphosphate carboxylase (known also as rubisco, RuBP). This is vital for the first major step when carbon is taken from carbon dioxide, fixed and becomes part of biological activity compared with its previous state as an atmospheric gas. Rubisco is the most abundant protein on earth and it is vital for the survival of all life.

Photosynthesis in those plants coming from the temperate geographical regions, for example Northern Europe, forms initially compounds containing three carbon atoms; this is known as C3 photosynthesis. These first carbon compounds are used in manufacturing the sugars fructose and glucose, which then combine into sucrose, thereafter forming starch and other compounds.

Plants such as cereals and grasses, which evolved in warmer temperate and tropical climates, such as Africa, survive in these arid areas by conserving water within their structures. This is achieved by closing the leaf pores (stomata) through which water exits and gases enter during daylight. These plants have evolved photosynthetic mechanisms which initially produce 4-carbon compounds; this is known as C4 photosynthesis. The energy requirements for C4 photosynthesis are greater than those of the C3 system, nonetheless this route is advantageous for plants which grow in hotter climates and are capable of utilising sunlight of greater intensities.

1.2 Using energy (respiration)

Respiration is the release of energy from the products of photosynthesis in the presence of oxygen and its use for physical, chemical and biological work by plants and animals (see Figure 1.9).

Respiration reverses photosynthesis by breaking down complex substances into simpler ones. This process is achieved as follows:

Sugar + oxygen →→→ carbon dioxide + water + energy release

Respiration involving oxygen is known as aerobic; there are circumstances where respiration may take place in the absence of oxygen; this is termed anaerobic. This latter happens with some microbes particularly those capable of living in hot water springs.

During daylight in favourable circumstances the rate of photosynthesis may exceed that of respiration in plants. In that case plants use up all the carbon dioxide formed from respiration and take in more from the atmosphere. The oxygen which released by photosynthesis is utilised for respiration in plants during daylight. There are circumstances when the light intensity is such that the rates of photosynthesis and respiration are equally balanced; this is known as the compensation point.

Fortunately, photosynthesising plants normally form an excess of oxygen and that surplus is released into the atmosphere. This provides supplies of oxygen which animals use. At night, photosynthesis ceases and plants are net users of atmospheric oxygen for respiration.

Aerobic respiration forms simpler compounds from more complex ones in a break-down process known as catabolism. This is where carbohydrates are broken down into sugars, proteins into amino acids and lipids into fatty acids which are then further degraded, releasing energy. Within each cell, respiration takes place in three stages.

1. Stage one (catabolism) takes place in the cytoplasm, which is the semi-fluid content which fills cells and within which are held other structures known as organelles each of which has specialist functions. During respiration complex molecules enter particular organelles known as mitochondria which are the power-houses in both plant and animal cells.
2. Stage two takes place in the mitochondria where enzymes dismantle the links between individual atoms, and that releases stored energy.
3. Stage three is the drawing-off of the released energy within the mitochondria by a process known as the Krebs cycle which has nine sequential steps (Hans Krebs was a German born British biologist who in 1937 unravelled this whole process).

The first three steps in the Krebs cycle prepare complex compounds for energy extraction, a process of priming, followed by six reaction steps forming the energy carrier molecules. If energy were produced in fewer steps, it would release heat, which would be both wasteful and potentially damaging for plants.

Once the energy is released, it is pumped out of the mitochondria in a series of steps, known as a transfer chain, involving a sequence of five protein molecules. Eventually, this makes the energy available for work in the form of growth, development, reproduction, and storage throughout the plant.

1.3 Moving goods around (transport)

Transport is the moving of nutrients and water from the roots into stems, leaves and flowers and the passage of the products of photosynthesis from leaves into the remainder of plants via the vascular system.

Transport in plants takes place in two directions in the vascular tissues present in the stems and roots, this consist of a series of tubes stretching from the root tips to the tops of plants, see Figure 1.10.

1. Firstly, the carbohydrates and other organic substances manufactured by photosynthesis must be moved downwards in solution into other parts of the plant. Depending on the position of leaves phloem transport may move compound upwards for example into fruit.
2. Secondly, water and nutrients must be moved up in plants from the roots, where they have been taken in from the soil or other substrates to the remainder of the plant and especially the leaves.

Downward and upward transport takes place in tubes of living cells, known as the phloem, while upward movement happens in tubes composed of inert cells known as the xylem. Combined, these two types of cell form the vascular tissues which are found either in the centre of plant roots and stems or in some plants scattered through the tissues. Eventually, the vascular tissues form the veins of leaves, flowers and fruits.

The xylem cells act like pipes extending from the roots to the highest shoot tips. Water and nutrients are drawn up these pipes by a process of evapo-transpiration. Water has a physical property whereby there is strong bonding between the molecules, this increases in inverse proportion to the diameter of the tubes through which it is passing. Consequently, the very narrow diameter of xylem tubes enables the movement of water to the tops of even very tall forest trees from their roots, for example the Giant Arborvitae (*Thuja plicata*) which originates from the west coast of North America and can reach 70 m in height.

As water is lost from leaves, a physical suction force is created at the top of plants which draws the water upwards. Quite literally, water is sucked upwards in plants by transpiration from the leaves as water exits through the stomata. Water flows into the roots through the root hairs, which are tiny single cells growing from the main roots. These cells are filled with water since there is a stronger solution of nutrients in the cells compared with the surrounding soil or artificial substrates. That disbalance in concentrations causes the physical flow of water into root hairs, a process termed osmosis. Minerals move into the roots along with water and are then transferred actively using energy released by respiration. This process moves water and minerals into the xylem. Thereafter, they are transported upwards through the stem into leaves, flowers and fruits where they can be used in cell development, growth, reproduction, photosynthesis and respiration.

Most of the water taken in at the roots is eventually lost through transpiration out through the stomata. Water passes from the thin-walled cells in the leaves into air-filled spaces beneath the stomata and from there it is evaporated. This process is driven by the temperature of the atmosphere outside the leaves, i.e. by heat from the sun or artificial sources.

The driving force for transport in the phloem is a ***gradient of hydrostatic pressure***. As long as sucrose is continually added at the source and removed at the sink the flow of sucrose will continue in this direction

1 At the source, sucrose is actively loaded into the phloem against a concentration gradient using energy in the form of ATP. The high concentration of sucrose means that water enters the phloem from the xylem by osmosis setting up a high turgor pressure

2 The water flows through the phloem under this pressure carrying the sucrose with it

3 At sink the sucrose is unloaded so there is a lower concentration and therefore a lower turgor pressure than at the source. This means the water is forced out of the phloem rather than entering by osmosis

4 Water is recirculated back up the xylem

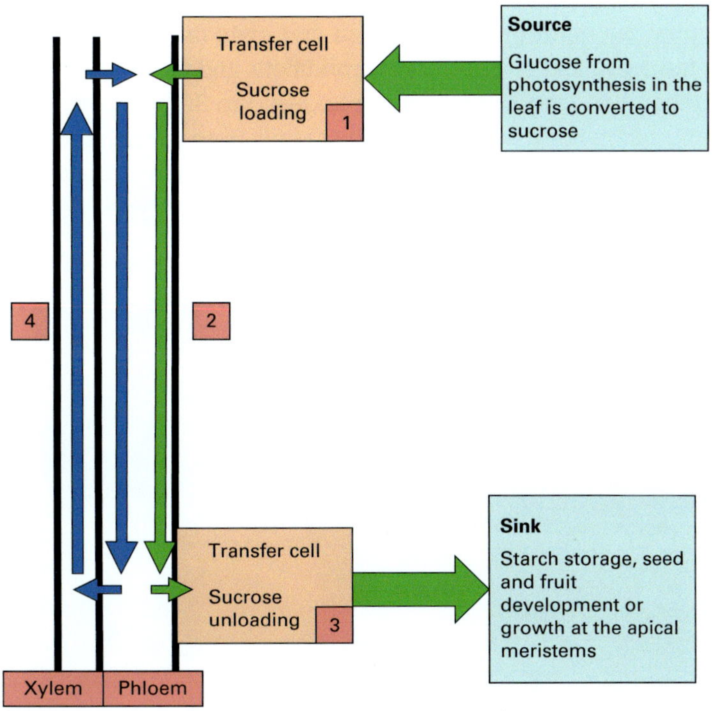

Figure 1.10 *Transport services*

Carbohydrates and other complex compounds manufactured in the leaves and other green organs are transported dissolved in water through the phloem tissues which are adjacent to the xylem. This transport process is known as translocation. In this process materials are moved from their source where they are made to a sink where they are utilised (see Underpinning knowledge 4.2, 5.2 and 7.2). Sinks are found predominantly at the tips of roots and stems and in developing flowers or fruit. Loading and unloading materials to and from the phloem are energy-consuming processes. There are specialised cells known as transfer cells at the source and sink through which loading and unloading takes place.

1.4 Getting bigger and stronger (growth)

Growth is an increase in the size, or biomass, of plants starting from the germination of seeds whereby the roots, stems, leaves, flowers and fruit increase physically in proportion and or in weight (see Figure 1.11).

Plant growth results from division within self-replicating layers of cells which are found at the tips of roots and stems known as meristems. These are constantly producing newly formed cells which then differentiate into a range of types depending on the function they will ultimately fulfil. These vary from those with thickened woody cell walls which make up the central support system to very thin-walled cells in the leaves which evaporate water and take in carbon dioxide and oxygen, supplying photosynthesis and respiration, respectively.

As plants grow, the number, location, size and even the structure of roots and leaves are determined by the prevailing environmental conditions and the genetic information passed from generation to generation (see Underpinning knowledge 3.1). Growth is controlled directly by growth-regulating substances, hormones, which result from the genetic constitution of the plant. Consequently, growth and development result from the interaction between inherited characteristics (traits) and the environment in what is termed the "genotype × environment interaction".

Figure 1.11 *A comparison of the growth stages of brassica seed, a seedling and a green broccoli head*

The visible plant body that results from this interaction is termed a "phenotype"; this is an expression of the combined effects of the genes inherited from parents and their interactions with the surrounding environment.

Research shows that there are strong parallels between the way in which growth and development are controlled in both animals and plants. This is because many of the basic chemical processes in plant and animal cells are closely similar, for example the use of energy through respiration.

Herbaceous ornamentals and most vegetable plants are composed of unthickened primary growth. Their cells are bounded by relatively thin-walled cells. Trees and shrubs produce secondary growth in which the cell walls become thickened by the accumulation of a woody material known as lignin. This provides the robust strengthening needed for life expectancies of several years or decades, and in the case of some forest trees for centuries.

"Herbaceous" is the term used when the leaves and stems either die completely after flowering and fruiting or the plant sheds the aerial organs, leaving a root system which may or may not survive winter conditions (see Chapter 7).

Plant cells can reformat themselves, developing into any of the organs present in the entire organism; this property is known as totipotency. Research with isolated cells grown in sterile culture showed that cells would initially form a mass of undifferentiated cells, or callus (see Chapter 8). Provision of suitable growth-regulator substances (hormones) encourages differentiation of callus into cells with specialised properties capable of forming roots, stems, leaves and fruits. This is the basis for the "micropropagation industry" which now worldwide reproduces huge numbers of herbaceous and woody plants commercially each year, some of which eventually are used in gardens (see Chapter 8). Under natural conditions, this is also the basis for vegetative propagation using parts of plants such as potato tubers, onion bulbs, vegetables, soft fruit, flower bulbs, and herbaceous plants, and propagation processes as discussed in Chapters 3, 4, 5, 6, 7 and 8.

Achievements: have you understood and learnt?

At the end of the chapter, you should be able to:

1. Understand that water, air, warmth, light, nutrients and substrates are essential for plant growth and reproduction.
2. Know that removing one of these factors prevents plant growth.
3. Understand that air contains oxygen and carbon dioxide both of which are needed for plant growth and reproduction.
4. Describe simple tests which identify the factors affecting plant growth and development, the need for nutrients and carbon dioxide and oxygen.
5. Understand that plants use carbon dioxide in the process of photosynthesis which builds sources of energy.
6. Understand that plants break down complex carbon compounds into simpler ones by the process of respiration and in so doing release energy used in doing work.

7. Describe the reliance of animal life on the photosynthetic ability of plants.
8. Understand that plants are able to transport water and nutrients from their roots to the uppermost shoot tip in the xylem cells.
9. Understand that the products of photosynthesis are distributed downwards and upwards in plants inside phloem tissues.
10. Know that all plant cells are capable of developing into any other cell, a property known as totipotency.
11. Describe how herbaceous and vegetable plants develop as the result of primary cell development.

Definition of terms: *Describe* is able to show why a scientific principle is important; *Understand* applies scientific principles into the context of gardening; *Know* carries through scientific principles in practical gardening activities.

Chapter 2
Understanding soil by growing potatoes and onions

Introduction

This chapter introduces the properties and importance of soil and its use. Understanding soil and encouraging its fertility are fundamentally important for the enjoyment of gardening, whether this takes place in the garden's land, in raised beds or on an allotment (see Figure 2.1).

Explanations are contained in this chapter about how gardeners physically till and tend soil. There follow descriptions of soil use for two relatively basic and easily grown crops, potatoes and onion sets. Potatoes in particular are valuable as a "cleaning crop". Their dense canopy of foliage out-competes other plants, weeds, for light and other resources suppressing them. Onions grown from sets are a straightforward crop for new gardeners. But like other similar bulbous crops they require careful husbandry by the gardener in keeping them weeds-free. The leaves of onions produce a vertical canopy which unlike potatoes does not suppress weed competition. Consequently, they require weeding from time to time so that growth is unchecked. The growth of both potatoes and onions is heavily reliant on a continuous supply of nutrients from the soil requiring applications of fertilisers before planting and top-dressings during the growing season. Fertilisers are added as firstly base-dressings made before seed sowing or transplanting and secondly as top-dressings applied into the standing crop.

Figure 2.1 *A vegetable garden*

Figure 2.2 *Earthworms in healthy soil*

Figure 2.3 *Massed annual weeds*

Garden practices

a. Soil management

Soils and composts are living entities containing vast numbers of microbes. Most of these are beneficial for plants and live in close association with the roots. Benefits flow from the microbes to the plant roots and also in the reverse direction. Soils are said to be healthy as the scale of this relationship increases. Soils and possibly composts also contain some microbes which are damaging for plants causing their diseases. Beneficial microbes help to keep these unwanted causes of disease in check in a process known as biological control (see Underpinning knowledge 4.3). Plant roots penetrate into soils and composts from which they draw water and nutrition which, as explained in Chapter 1, are vitally important for plant growth and reproduction. Additionally, soils and composts provide physical anchorage and stability for growing roots. Soil-borne animals, especially earthworms (see Figure 2.2), help improve the physical and chemical nature of soil by channelling, burrowing and digesting.

This improves soil health and benefits plant roots. Many insects, like the microbes, live in the soil and are beneficial for garden plants. There are some that are pests requiring control, such as slugs, snails and wireworms.

b. Weed control

Weeds may be considered as "plants growing in the wrong place"; in the garden they are generally wild native plants or imported aliens. These plants thrive and grow luxuriantly as a result of high levels of soil fertility and possibly a lack of competitors. Large volumes of weed seeds are present in soils, this is known as the 'seed bank'. These will produce weed plants which compete with crops reducing their productivity. Controlling germinating and seedling weeds is of major importance since this prevents competition with existing crops and reduces the seed bank of the soil minimising problems for future years (see Figure 2.3).

Once weeds become established and flower, producing their own seed, control becomes far more difficult, as described in the adage "one year's seeding is seven year's weeding".

Weeds which grow quickly producing seeds and dying within one season are known as "annuals" and some, for example groundsel (*Senecio vulgaris*), produce several generations in one season (see Figure 4.63). These are controllable at an early stage in their growth by hoeing or using specialised chemicals. Others are "perennials", for example couch grass (*Agropyron repens*) and are more of a problem since they live for several years and produce very deeply penetrating roots (see Figure 2.4).

Figure 2.4 *Perennial weed, couch grass (*Agropyron repens*) competing with radish plants*

Control of perennial weeds is more difficult and requires long and careful digging of the land and the removal of the roots. The problems which are associated with perennial weeds are most acute on land which has been left uncultivated for several seasons, for example, when the tenancy of an allotment has lapsed. Faced with this problem the most effective remedy is covering the soil with black polythene sheets in the early spring. This has the effect of encouraging perennial weed growth while at the same time weakening it because plants are unable to benefit from light. Once the polythene sheets are removed, then digging out the weakened roots becomes easier and more effective; also, they are more susceptible to chemical weed-killers.

c. Selecting and preparing land

The most satisfying returns for gardeners come from selecting land for growing vegetables and fruit which is as level, stone-free and as weed-free as possible. This is a counsel of perfection and may not be easily achieved, certainly when beginning a new garden. The garden may be part of a newly-erected house or estate in which case it is likely that the builders will have left legacies of buried bricks, stones or cement which require removal. An allotment tenancy may bring problems caused by underuse and resultant weed infestation. In both cases the gardener may be encouraged into the use of raised beds filled with proprietary compost. While this may offer good germination and growth initially, these structures demand continual care, preventing drying out as the plants reach maturity.

As seen in Chapter 1, the soil supplies plant roots with water, air, anchorage and nutrients. Maximising these benefits requires that the soil has high fertility and is in good health; healthy soils are those containing large numbers of beneficial microbes and abundant animals such as earthworms. Achieving fertility results in most cases from the addition of considerable quantities of organic matter. The best and possibly most difficult organic manure to obtain is well-rotted horse manure (see Figures 2.5 and 2.6).

Figure 2.5 *Handling animal manure*

Figure 2.6 *Animal manure close up*

Figure 2.7 *Digging tools: spade, digging fork, manure fork (left to right), in front is a builder's trowel for cleaning the other tools*

Other animal manures are effective means of improving soil structure but tend to contain larger quantities of soluble nitrogen and this can cause over-rapid and lush plant growth, which can be associated with increasing disease incidence. Alternatively, there are several proprietary brands of organic compost which can be incorporated.

d. Digging soil

Manure or well decomposed compost is added into the soil by digging which is best achieved in late autumn or early winter, generally between November and the end of January. Timing depends on the weather conditions as climate change is extending autumn seasons which are now warmer for longer and also wetter. Digging very wet soil is difficult and laborious. Wet soil particles clog together and cling onto the spade blade. Walking on very wet soil causes compaction destroying its structure and reducing aeration.

Digging requires tools (see Figure 2.7), these include a digging spade and fork and preferably a manure fork which has curved tines which will pick up equalised amounts of manure from a heap.

A good digging spade is one of the gardener's biggest joys and most precious tools. Do not buy on price! A cheap spade will be badly balanced and made of poor-quality steel, making for tiring work and a tool that wears badly. Buy a spade with a "D"-shaped handle, wooden shaft and stainless-steel blade, which can be sharpened from time to time. When considering which spade to buy, balance the tool in your right hand (or left if you are left-handed), with your hand placed at the point where the top of the collar of the blade fits into the handle (see Figure P1.2 in the Preamble). At that point, the tool should balance on your hand and seem almost weightless. The result is that when you are digging, you will be lifting the weight of the soil on the blade, not the weight of the spade. Over years the spade's blade will need sharpening for ease of cutting into the soil. This is most easily done with a sharpening stone, which should be drawn repeatedly across the inside edge of the blade while the spade is held upright and resting on a wooden bench or plastic box.

e. Digging processes

Wearing a good pair of Wellington boots and thick socks makes digging a more pleasant experience in December and January. The essence of digging is moving and inverting soil, incorporating organic matter and burying or removing weeds in the least arduous manner. This is accomplished by firstly taking out a trench along one of the edges of the plot (see Figure 2.8).

Hold your spade in both hands, one on the handle and one placed about the collar of the blade where it joins the wooden shaft. Determine where the first spade full of soil is to be removed, chop down into the soil along the front edge and then either side of the piece of soil to be removed. The full length of the spade's blade is termed one "spit". One spit is the depth of soil cut by a spade; this is a 16th-century Middle Dutch term and is mentioned in the 1680s by John Evelyn in his *Directions for the Gardener.*

These cuts should be the width of the spade's blade. Then push the entire blade down on the back edge of the soil so that a square of soil has been marked. Now press down on the handle to lever the spit of soil, lift the spade-full of soil and spread it behind you onto the land which will be eventually dug. Bend from the knees avoiding placing strains on your back muscles. This process is then repeated all the way across the piece of land until a trench which is the full length of the ground has been established. This trench should be about as deep as the length of the spade's blade (one spit) (see Figures 2.9, 2.10 and 2.11).

Now with the manure fork spread compost or manure evenly into the bottom of the trench to a depth of about 5 to 6 cm. If there are lumps in the manure, these should be broken up with the edge of the manure fork's tines. Clean the spade's blade with an old builder's trowel removing any clinging soil. A dirty spade blade makes digging much more laborious and tiring and less effective (see Figures 2.12, 2.13 and 2.14).

Figure 2.8 *A first trench*

Figure 2.9 *Laying manure in the first trench*

Figure 2.10 *Laying manure into the first trench with help from a robin*

Figure 2.11 *Cutting the first spade full of soil (spit)*

Figure 2.12 *Turning and laying the spade full of soil*

Follow this procedure at the end of each row of digging.

Some gardeners find a garden line marking the breadth of the next run of digging helpful (see Figure 2.15).

Figure 2.13 *Soil clinging to a spade blade*

Figure 2.14 *Cleaned spade blade*

Figure 2.15 *Garden line as a guide for straight rows*

This equalises each spit keeping the size of spades-full of soil even. Garden lines are essentially a piece of string as long as your plot is wide tied between two sticks. More sophisticated versions are available consisting of a metal pin and a rotating bobbin onto which the string is wound. The pin is inserted at one end of the proposed row and then the string laid across the plot and pulled tight on the bobbin giving a straight line along which to work.

Stand squarely in front to the first trench which is now filled with manure. Mark out one spit of soil chopping down with the spade blade on either side. As previously suggested, press the full length of the spade down at the back of the spit. Raise a fully loaded spade of soil and invert it into the trench. The art here is in fully burying all the manure and any weeds which may be growing on that spade full of soil. Repeat this process right across the piece of land, ensuring that the inverted spits of soil are even and regular. If the land is wet, break up the surface of the inverted soil by pushing the spade blade into it to a depth of 4 or 5 cm. This allows the entry of frost more quickly which is beneficial in refining the soil surface for future use (see Figure 2.16).

You will have created another trench into which manure can be spread with the fork and then covered with soil dug from the next run across the width of the land. Slowly and surely, you will see a very satisfying expanding area of dug soil stretching out in front of you which should be

Figure 2.16 *Well-frosted soil*

completely free of any evidence of weeds and with the manure entirely buried beneath the surface of the inverted soil. Establish a rhythm and pace for your digging actions with which you feel comfortable, this makes the entire process easier, more energy-efficient and far less tiring. Ultimately, there are few sights in gardening more satisfying than seeing the finally finished area of dug ground in front of you.

f. Follow-up husbandry

Dug land should be left undisturbed for a couple of winter months in the anticipation that it will become deeply frosted. Frost helps break down the dug soil preparing for a fine tilth which will be invaluable for sowing seed and transplanting newly raised seedlings in the spring. Tilth comes from the Old English "tilthe" and is related to tillage or cultivation of soil; a fine tilth has fine aggregations of soil particles suitable for small-seeded crops.

This is achieved by the alternating freezing and thawing of moisture within the soil, which shatters soil clods and produces an even tilth. Gardeners should repeat the digging and manure incorporation processes annually as part of their seasonal rhythm. This will gradually increase soil fertility and encourage increasing populations of earthworms.

Noticeably, crops grown in land which receives manure each year will gradually require less watering during periods of dry weather. This is because opportunities have been created for deeper rooting and hence increased access to soil water reserves. The soil gains an increasingly retentive structure, or quality, which holds more water and air which roots can draw upon. Soil quality refers to its ability to produce good tilths for sowing and planting, and ultimately, high-quality and nutritious crops.

g. Liming

Once land has been manured and dug over, liming may be needed. Lime further improves soil structure and the availability of nutrients. The majority of limes supply calcium, which is required for plant growth and the strengthening of cell walls. Lime also reduces soil acidity and increases alkalinity values (see Underpinning knowledge 2.4). That makes soil nutrients more readily available for roots. Over time most soils naturally become more acidic. This is because rainwater washes weak acids, such as carbonic acid, from the atmosphere in to them. The need for lime can be tested using proprietary kits which estimate "soil pH" and are obtainable from gardening centres or by mail order (see Underpinning knowledge 2.4).

Small quantities of lime can be bought from garden centres and are spread according to the instructions given by the manufacturer which will be printed on the bag. Lime is relatively light weight and can be caught by wind gusts. Consequently, spreading is best achieved on still days with any breeze blowing from behind you onto the plot. Bend down and spread the lime closely to the soil surface and distribute the powder as evenly as possible. Lime can take several months before breaking down sufficiently, altering soil pH and making calcium available; consequently, applications should be made in winter after the land has been dug.

New gardens in developing estates may have poor-quality soil structure and fertility. This results from buried rubble which is often highly alkaline and where soils have been heavily compacted by trafficking during the building work. Digging the entire site, incorporating as much organic matter as possible, is the best policy and will yield increasing rewards for the future. In these circumstances, do not wait for autumn; this task should be done as soon as the owner gains possession of the land. In the process of digging, all rubble can be removed along with weeds, especially those of a perennial nature such as dandelions (*Taraxacum officinale*), docks (*Rumex* spp.) and couch grass (*A. repens*). The site should then be sown with grass seed and that crop retained for an entire season. For this task a "general purpose" grass seed mixture is fully adequate as it will contain mostly perennial ryegrass (*Lolium perenne*). The grass roots will help develop soil fertility and structure by binding soil particles into a mat of fibrous roots. Once the grass has formed a temporary lawn, this can be dug and manure or compost incorporated. Ensure that the turfs of matted grass are fully inverted when digging; this encourages their breakdown, preventing re-emergence.

h. Using dug land

Late winter and early spring heralds the time when the garden will move into planting and production. This can start as early as the beginning of February in favoured southern areas or as late as mid-April or even May in northerly parts, especially in Scotland. Gardens which have been dug during the winter should by this time be showing signs of weathering as the spades full of soil will have disintegrated. Now is the time for secondary cultivations using a rake. Gardeners should provide themselves with a good-quality rake which has a head of about 30 to 40 cm width and tines that point downwards and are 8 to 10 cm long and has a long wooden handle. This implement should balance in the hand in a manner similar to that illustrated for the spade (see Figure P1.2). Gardeners will also require a good digging fork which will have the characteristics of the spade in terms of balance and comfort in use. While buying these tools, gardeners may also like to purchase a square-headed hoe. This will have a head which is a square of steel pointing down from the shaft and can be used in the seed-sowing operation.

i. Planting potato tubers (seed)

Potatoes are a popular crop with most gardeners, especially the very early crops of new potatoes which when gathered directly have a distinctly succulent flavour often far superior to those available in supermarkets or other shops. This is because the tubers are freshly dug, cooked and eaten very quickly after harvest. Potatoes are classified into 'early', 'mid-season' and 'late (maincrop)' groups of cultivars (varieties). The correct term for artificially created varieties is cultivar but "variety" is retained in common use.

Early potatoes may be harvested from mid-June onwards, while mid-season varieties provide useable tubers from about mid-August into early September and the maincrop types are harvested and possibly stored in early October. These maturities vary with season and are later further north especially in Scotland.

Greatest return for effort for the gardener probably comes from the early types because there is now such a huge over-supply of maincrop tubers on the market that the prices are relatively low. But for those gardeners growing each of these groups of potatoes the practices are very similar.

j. Seed potatoes

Purchase a supply of 'seed' tubers either from a seed supplier or from the garden centre in early February in the south or latter in the north (see Figure 2.17).

A popular early and quickly maturing type is the variety 'Swift'. Others of this well-flavoured type include: 'Casablanca', 'Maris Bard', 'Duke of York' and 'Sharpe's Express'. Second early varieties include: 'Athlete', 'Cerisa', 'Charlotte', 'Jazzy' and 'Orchestra'. Late maincrop varieties include: 'Alouette', 'Cara', 'Carolus', 'Manitou' and 'Nicola' with particular heavy yields of baking-size tubers from the well-tested standards like 'Maris Piper' and 'King Edward'.

Botanically, potato tubers are swollen stems and as a result each have buds on the surface and the first task is to encourage these into active growth. Buds on potato tubers are also known as "eyes".

Place a sheet of old newspaper into the bottom of a seed tray and arrange the potato tubers in it. Give space between the tubers and look to see that there are no injuries or disease patches evident on their surfaces. Removing some of the smaller buds by rubbing them with a forefinger and thumb encourages stronger growth by those that remain. Careful examination of each tuber will show a small scar on one end, this is where the tuber was attached to the mother plant. The other, rose-end, is where the most vigorous buds will be found.

Figure 2.17 *Seed potatoes on sale*

Stand each tuber such that the greatest number of buds are upper most. Place the tray in a darkened, cool, airy place such as a frost-proof garage. Ensure that the tray cannot be reached by vermin such as mice as the tubers will provide a welcome winter meal for such guests. Keep a watchful eye on the tubers for such attacks and for the development (chitting) of shoots on the tubers (see Figures 2.18 and 2.19).

From mid-March in the south and into mid-April further north, the tubers can be planted into the garden. There is a balance between encouraging potatoes into early growth in the

Figure 2.18 *Well-sprouted (chitted) potato seed (tubers)*

Figure 2.19 *Close-up of seed potato sprouts, ready for planting*

garden and avoiding damage to developing foliage caused by frost. Potatoes are very sensitive to cold weather because they originated from warm temperate, frost-free areas of South America. The gardener will accumulate local knowledge from the garden over several seasons as to when planting can safely take place and initially useful advice may be obtained from neighbours.

k. Planting husbandry

Break down an area of the vegetable garden which has been previously dug and left for winter frosting with a rake. It should now be in an easily worked condition and provide ground which can be planted with potatoes. Soil that is easily worked is termed "friable", it breaks down into small crumbs of soil, about 0.5 to 1.0 cm diameter or less, which do not stick to the tools.

When planting the tubers, place a walking board onto the vegetable plot. Walking boards spread the gardener's weight reducing compaction of the soil underneath. Mark out a straight row across the garden with a line for guidance in positioning the potatoes. Dig out a deep trench (20 to 25 cm deep) alongside the line with the spade carefully piling the soil in a ridge alongside the trench (see Figure 2.20).

Figure 2.20 *Excavate a trench and add granular fertiliser*

Figure 2.21 *Mix fertiliser into the bottom of the trench*

Figure 2.22 *Lay old newspaper in the bottom of the trench*

Figure 2.23 *Wet the old newspaper*

This is best done when the soil is dry and easily worked. Sprinkle "GrowMore"-type fertiliser (see Preamble) into the bottom of the trench at 30 g per square metre and work it in lightly with a fork (see Figure 2.21).

Place a layer of old newspapers in the trench and water these well. This provides a reservoir of moisture on which the developing tubers can draw (see Figures 2.22 and 2.23).

Place a thin layer of soil on top of the wet papers and then carefully stand each potato tuber in the trench spaced at 20 to 25 cm apart (see Figure 2.24).

Cover the tubers with soil taken from that placed in the ridge alongside the trench. Mound up the soil with more taken from alongside the ridge where the potatoes are now planted (Figure 2.25) and clean your tools (Figure 2.26). This means that the tubers are buried under about 20 to 30 cm of soil.

Figure 2.24 *Cover the newspaper with soil and evenly space out the seed potato tubers in the trench*

Figure 2.25 *Cover the seed potato tubers with soil and mound up into a ridge*

I. Watch for tender growth

Monitor the potato ridge carefully for signs of emerging green shoots after 3 to 4 weeks. The exact time of emergence depends on weather conditions and especially soil temperature as a period of warm weather encourages more rapid emergence (see Figure 2.27).

The first green shoots should be covered by drawing more soil onto the ridge with a broad-bladed hoe. But there is a limit to the amount of soil which can be drawn onto the ridge. The most effective frost prevention is using either a row of glass barn-cloches, polythene cloches or lengths of horticultural fleece placed carefully over the ridge. Alternatively, if these are not available then cover with wet newspaper or an old counterpane for the night and remove these as soon as the frost clears the next morning, so that growth is not damaged (compare Figures 2.28 and 2.29).

Figure 2.26 *Clean your tools*

Figure 2.27 *Emerging potato shoots*

Figure 2.28 *Well-developed young potato plant*

Figure 2.29 *Potato plants suffering from frost damage*

Figure 2.30 *Potato in flower*

Follow the progress of emerging potato shoots and consult the local weather forecasts on Google, appropriate Apps or on radio, television or in daily newspapers for warnings of impending low overnight temperatures.

m. Potato crop husbandry

Potatoes are gross feeders and withdraw large amounts of water and nutrients from the soil especially when initiating and forming succulent new tubers. If there is an extended period of dry weather, the potato rows should be watered well. The early varieties flower in the first weeks of June and this is good indication that tubers are also swelling (see Figure 2.30).

Feed the plants with diluted tomato feed or a proprietary liquid seaweed extract as that will boost growth and increase the quality of the tubers. Early-season crops should be given fertiliser with a nutrient ratio of 1:2:1 (N:P:K), while later-season crops benefit from fertilisers with the ratio 1:2:2 (N:P:K). These figures describe the relative amounts of the nutrients nitrogen, phosphorus and potassium in the fertiliser and are declared on the containers. The meaning of fertiliser nutrient ratios is discussed in Underpinning knowledge 2.4.

Early varieties are grown for the benefit of relatively small, succulent tubers. It will be worthwhile digging up a test plant, determining if there are sufficient new tubers ready for the cooking pot. These should be harvested in the immature stage when the skins can be rubbed off with the thumb (see Figures 2.31 and 2.32).

If the gardener wants larger quantities of storable tubers, then it is advisable to plant main crop varieties. These will be left in the ground until the foliage (haulm) has matured and senesced in early to mid-September. At that stage, the tubers can be lifted and placed in a tray in a cool, darkened place but well away from predations by mice. That allows the tuber skins to dry and mature. Once they are firm, then the tubers can be stored in hessian sacks again well away from vermin. Do not attempt to store any damaged or diseased tubers as these will rot and infect surrounding ones. The mid-season cultivars follow-on from the early-season varieties giving a sequence of tubers from June through to September. These should be harvested as required for the kitchen they will not store satisfactorily beyond two or three weeks. Any tubers that are stored should be kept away from light as this encourages greening by chlorophyll production in the skins which imparts a bitter taste and it is not advisable for human consumption and especially not by pregnant women. A sequence of potatoes from early types through mid-season to main crops will require a considerable amount of land. In a small domestic garden, this is not a sensible use of space.

n. Disease control

Potato crops growing beyond mid-June will require careful monitoring for the incidence of potato blight disease (*Phytophthora infestans*). This microbial disease can very quickly devastate crops (see Figure 2.33).

At the first signs that the foliage is showing blight symptoms, a fungicide should be applied in accordance with the manufacturer's recommendations. The foliage left after harvesting potatoes should be disposed of as green municipal waste and not composted in the garden. This is

Figure 2.31 *Maturing early potato crop 'Lady Rosetta'*

Figure 2.32 *Freshly dug early potatoes*

Figure 2.33 *Damage caused by potato blight (*Phytophthora infestans*)*

Figure 2.34 *Onion bulbs (sets)*

because infections of potato blight may remain viable on the haulm and can be carried over into the next season. Potato stems and foliage are collectively called "haulm".

Tubers in the ground in September are prone to invasion from wireworms or slugs. These animals bore into the tubers, making them inedible and encouraging rotting caused by various soil-inhabiting fungi and bacteria. For this reason, it is advisable to lift the tubers as early as possible in September. Slugs can be a considerable menace because they will also browse on potato shoots. Applications of biological control are helpful using preparations of predatory eelworms (such as *Phasmarhabditis hermaphrodita*) which parasitize slugs. Using this technique also helps preserve the garden's bird population since they can then eat the dead slugs with impunity.

o. Alternative husbandry

Gardeners may grow potatoes in barrels or growbags as means of obtaining earlier or later crops. In either case, the tubers are planted in nutrient-rich composts and can be protected from low temperatures by placing in a greenhouse. This means that crops of succulent new potatoes may be harvested either very early in the season or very late perhaps providing an additional vegetable at the Christmas table.

p. Onion sets

Onion sets are bulbs which have been grown by specialist commercial producers during the previous season. These bulbs are then used both by growers and gardeners as a means for gaining earliness in the maturity of their crops. Normally, sets come into maturity in July and August well in advance of crops sown from onion seed. Bags of sets can be purchased either directly from seed companies or in the garden centre. Bags sold to gardeners usually contain about 20 sets (see Figure 2.34).

When purchasing these, it is advisable to gently squeeze each bulb to ensure that it is robust and not affected by onion rot (*Sclerotinia allii*) fungal infection.

Once purchased, remove the onion sets from their container and place them into a seed tray which has been lined with old newspaper, in a similar manner to potato tubers (see Figure 2.18).

Place the tray in a cool dark place such as a garage but well away from possible predations from mice. Normally, these rodents are averse to eating such foodstuffs but in a hard winter they may find them attractive. The sets are suitable for planting when green shoots have just started emerging from the bulb at which time root initials may be visible. A suitable variety is 'Hercules', this will provide mature bulbs of about 6 to 8 cm in diameter which may be used directly in the kitchen or stored. Those gardeners who fortunately live in locations where the soil temperatures rise rapidly in early spring may find that onion sets can be planted directly without first starting growth under protection.

q. Starting onion growth

A growth advantage may be gained by starting onion sets in a greenhouse. A plastic tray constructed with a series of modular units should be filled with proprietary seed compost obtained from the garden centre or other sources. Single onion sets are pushed gently into the surface of each module and the whole is placed on the greenhouse staging (see Figure 2.35).

Water well and monitor for the development of green shoots and roots emerging from the bases of the modules. Once growth is sufficient, then the sets may be carefully planted directly into the garden (see Figures 2.36 and 2.37).

r. Planting onion sets

Planting onion sets either directly from a packet or as growing plants follows similar procedures. Once your piece of ground intended for the onion sets has been well raked to a tilth, lay the walking board across the plot in parallel with where you are intending to plant the sets. Land which has been recently acquired may need the addition of fertilisers, in that case mix into the soil a dressing of general "GrowMore" fertiliser at the rate of 30g per square metre. Where there has been continual incorporation of manure and hence highly fertile may not require fertilisers prior to planting. Layout a straight row, squarely placed across the plot using a garden line. Using the hoe turned so that the edge of the square blade provides a 'V' shape draw it across the plot alongside the string to a depth of 5 to 8 cm. The soil will mound up on either side of this small

Figure 2.35 *Onion bulbs placed in a modular tray*

Figure 2.36 *Onion plants ready for transplanting into the vegetable garden*

Figure 2.37 *Modular tray filled with onion plants ready for transplanting*

Figure 2.38 *Onions planted out along a garden line*

trench. Take each bulb from the seed tray where they have been chitted and gently place it into the trench. Look carefully at each bulb and you will see that each has a definite bottom which shows signs of producing root initials and a pointed top where new leaves are emerging.

When planting modules, take out a trowel full of soil and stand the growing module into the hole. Place soil around the developing plant. Space the individual dry bulbs or modules at 10 cm distances from each other along the trench; the distance can be measured using a piece of bamboo cane cut to the correct length. Once all the bulbs or modules are safely placed in the trench, draw soil from the trench ridges onto the bulbs and refill the "V" shape. The soil should just cover the bulb, leaving the green shoots fully visible (see Figure 2.38).

The bulbs can be settled into their positions by watering gently from a can fitted with a coarse rose. Remove the walking board and lightly fork over its impression on the soil. This last action should become routine each time that the gardener has stood on the vegetable garden's soil. Forking over helps avoid soil compaction, encouraging aeration and the flow of water into soil, each of which enhances soil fertility.

s. Onion husbandry

Onion sets have few enemies although gardeners may find that birds especially pigeons can pull-out the sets largely for devilment as they will not eat the bulbs. Pets such as cats and dogs may dig around in the loose cultivated soil. If there are such problems, it would be advisable to place a wire netting cage over the row until the bulbs have developed sufficiently strong root systems which stabilising them in the soil. At that point, the cage can be removed. Onions are gross users of soil nutrients. Consequently, applications of a suitable liquid feed should be made at intervals during their growing season. Liquid feeds sold for tomato culture are suitable for this purpose. Dilute the concentrated feed at the rates recommended for tomatoes for this purpose and apply about every three weeks. Onions grown on land which has been well treated with farmyard manure for several years will probably contain sufficient residual nitrogen, so that liquid feeds which preponderantly contain phosphorus and potassium are preferable. Excess nitrogen will encourage the incidence of diseases such as onion neck rot (*B. allii*) (see Figure 2.54). Alternatively, use liquid seaweed extracts, which have the added value of improving bulb quality and freedom from diseases.

t. Onion growth

The onion sets will produce substantial leaf growth. But since the leaves are cylindrical in shape and tend to grow directly upwards the leaves do not cover the space between them as is the case with crops with spreading canopies such as potatoes. This means that weeds can grow between the onion plants competing for light and nutrients. Weeds will also produce a moist environment, which is conducive for some of the leaf-invading plant diseases such as downy mildew (*Peronospora destructor*) or rust (*Puccinia allii*) which affect onions. Weeds should therefore, be removed from the onion bed regularly before they have been able to damage the growth of the crop. If there is a spell of hot, dry weather, then it is advisable to water the onions frequently, maintaining steady stress-free growth. Onions that suffer from water restriction tend to "bolt"; in other words, they produce a flower spike. This is not wanted as bulbs that flower produce small, poor-quality bulbs that will not keep for any length of time.

u. Harvesting onions

After about 2.5 to 3 months' growth, the new, much enlarged bulbs will be very obvious on the surface of the soil and should be 5 to 6 cm in diameter with a developing dry brown coat. The leaves will "fall" when they drop towards the ground; this is natural and normal because the plants have reached maturity and the leaves are senescing. Pull up the onion bulbs and leave them on the soil surface, where they can dry naturally if the weather is fine and dry and predictably stable for 7 to 10 days. If the weather is unpredictable and wet, lift the bulbs into a tray lined with old newspaper and place them in a dry garage where their senescence can be completed. Once the leaves are dried, remove them and residual soil from the bulbs and place the bulbs in a netting sack hung in the garage and use at will in your kitchen; well-dried bulbs will keep for several months (see Figure 2.39).

Figure 2.39 *Mature culinary onions for storage*

Underpinning knowledge

2.1 Knowing soil

Soil is a very thin layer of natural material into which plants root and from which they obtain their livelihoods, without which there would be no life on Earth. Healthy, good-quality soil is an essential pre-requisite for successful gardening. Understanding how this is achieved is an essential gardener's skill. The terms used for describing good-quality and healthy soils and composts include:

Fertility: this describes the capacity of soil for providing robust, high-quality plant growth. It embraces the nutrients present, the favourable biological components and physical properties whereby air and water enter and are retained in the soil encouraging growth. A fertile soil will have a structure made up of particles of differing sizes; these form channels in the soil whereby water drainage takes place. Free-running drainage from a fertile soil prevents waterlogging. Soils suffering from waterlogging are deficient in air, and as a consequence plant roots are deprived of oxygen. Roots require oxygen for respiration and water is the medium for all chemical reactions in root cells (as described in Underpinning knowledge 2.1, 2.2 and 2.3). The carbon dioxide produced by roots during respiration diffuses into the air or dissolves in excess water as it drains down through the soil.

Aeration: air in the soil is a vital component because all roots need oxygen and the quick removal of carbon dioxide. Effective aeration prevents the accumulation of toxic concentrations of carbon dioxide, which could asphyxiate roots. A well-structured soil permits a rapid interchange of oxygen and carbon dioxide between the aerial atmosphere and that within soils.

Tilth: soil with a high fertility will break down to finer particles this encourages the emergence of roots from young seedling which then start drawing in nutrients, oxygen and water. A finely structured tilth results from soil breaking down through the action of frost and a gardener's husbandry (see Figure 2.40).

This soil can be worked in preparation for sowing small seeds such as beetroot, carrots, onions or parsnips. These are often sown directly into the soil. If the soil is composed of large particles or clods, greater than 5 mm, small seeds germinate unevenly and the roots cannot penetrate effectively. This produces erratic germination and weak seedlings with resultant disappointing returns for the gardener. These unthrifty seedlings are prone to infection from soil-borne pathogenic fungi causing problems generally described as "damping-off" diseases, where the seedlings appear wet and waterlogged at the stem base and are often covered with a mat of white fungal growth (see Figure 2.41).

Figure 2.40 *Good friable well-structured soil tilth in the vegetable garden*

Moisture: soils must supply moisture for roots in adequate but not excessive quantities.

Where the soil is waterlogged, the roots become damaged, unable to take in oxygen and affected adversely by toxic carbon dioxide. Under these conditions, seedlings are susceptible to infection by soil-borne diseases which cause damping-off.

Figure 2.41 *Seedlings affected by damping-off disease caused by a range of "water moulds". Note the way in which the seedling stems are "pinched" at the base, this is a classical symptom*

Healthy soil is well structured, contains a high level of organic matter, large populations of beneficial microbes, and holds water supplies but drains adequately following rain or watering. An under-supply of water soon becomes evident as plants and more importantly seedlings begin wilting. Wilting must be remedied rapidly. Improving structure and increasing the organic-matter content of the soil raises fertility and is a longer-term solution for supplying adequate moisture. A fertile soil well supplied with organic matter will provide both the air and water required by plants more effectively and for a greater part of the growing season. Healthy soils are also well supplied with beneficial microbes which aid plant growth and help prevent soil-borne diseases.

Soil is developed either from parent rocks or by the deposition of organic matter, usually peat, over many centuries. There are two basic soil types, those with limited organic matter content (below 5 per cent) and those with very large amounts. These are termed "mineral" and "organic" soils, respectively.

Weathering

Over geological time parent rocks are broken down firstly by physical processes and then by chemical and biological actions, which are collectively known as weathering. The simple process of water washing over and through rocks breaks them down especially the softer forms such as chalk, limestone and sandstone. Water expands when it freezes forcing the break-up of rocks into boulders, which flood water then moves across the landscape. The action of boulders bouncing against each other breaks them further reducing their size to pebbles. Hot weather heats rocks expanding the outer layers causing them to peel away from those underneath. As the size of rock fragments is reduced, they are collected by wind and blasted against other rocky surfaces, further reducing the particle size and increasing the rate of degradation. Wind, water and ice (glaciers) transport large quantities of rock particles over very considerable distances. Glaciers, in particular, grind down substantial boulders, transporting the resultant materials slowly over many kilometres. Violent earth movements such as volcanoes, earthquakes and severe weather events such as hurricanes, tornadoes, tsunami and cyclones contribute further to the breakdown of rock, grinding the resultant particles and transporting them over long distances. Degraded rock is also weathered by chemical and biological processes. Rocks are dissolved naturally by carbonic, sulphuric and nitric acids. Carbonic acid is formed from atmospheric carbon dioxide. Sulphuric acid forms by the action of water on sulphates within the rocks and nitric acid results from the effects of electrical discharges, lightening, in the atmosphere which turns nitrogen gas into nitric and nitrous oxides. Pollution resulting from aerial discharges from factories and power stations may also be sources of sulphuric acid, known as "acid rain". Animals such as earthworms swallow fine rock particles digesting them with their gut acids. The resultant degraded material may either reflect in its composition the underlying rocks, or alternatively rock fragments are moved well away from their parental material and as a result bear little or no relationship to the layers on which they are deposited.

Soil transport

Soils themselves may be products of transportation by water, wind, in glaciers or down slopes by gravity (see Figures 2.42 and 2.43).

Most commonly, water transport produces rich alluvial soils such as those which supported the Egyptian civilisations for about 5,000 years along the River Nile flood plains. In the most recent Ice Age melting glaciers deposited mineral-rich soils over much of the northern hemisphere additionally winds produced layers of loess. Loess is a very fine-grained, silty, wind-blown deposit found in the valleys of the rivers Rhine and Mississippi and northern China, forming an estimated 10 per cent of the world's soils.

Gravity results in screes typical of mountainous regions, which are characteristically slowly colonised by alpine plants.

Figure 2.42 *Soil erosion by water, carriage of a flume of soil in water run-off*

Soil profiles

Digging a pit into soil as deeply as possible will reveal a soil profile and roots growing down and being supported (see Figure 2.44).

Figure 2.43 *Soil erosion by wind*

Figure 2.44 *Soil profile*

Soils that reflect the properties of parental rock which they overlay are composed of a series of layers or horizons. Accumulated directly on top of the soil is a layer of organic litter formed from leaves, animal manure, worm castes and other biological material, which is gradually and continuously decaying and uniting with the soil. The upper layer (horizon) is the fertile topsoil where plants obtain most of their nutrients and the preponderance of roots provide structural strength which supports aerial growth. Healthy topsoil is well aerated and should allow water percolation such that plants are able to gain the oxygen and water which they require (see Chapter 1). Below the topsoil are layers known as subsoils, some of these are thin comprising only a few centimetres in depth, while others can extend for more than a metre. Some soils are composed almost entirely of the products of decayed plant material and these are termed organic soils. They can be extremely fertile and high yielding, but require careful attention because seedlings and young plants in particular, may be exposed to droughts.

Soil structure

Fertile soils, topsoil, characteristically contain active biological life, ample nutrients and have darker colour compared with the subsoil. The term soil structure is used to describe its physical characteristics. Well-structured topsoil permits gas exchange with the aerial atmosphere supplying roots with oxygen and diffusing away carbon dioxide. They store sufficient water, which satisfies the requirements of roots but does not prevent water percolation when rain is heavy. This is achieved because the soil has an effective crumb structure. Individual soil particles aggregate together forming soil crumbs resulting from their chemical composition and adhesives produced when organic matter is degraded by soil microbes. Ideally fertile, friable soils have a crumb structure which encourages root growth and allows the percolation of water into the subsoil and dispersal away into catchment areas. Where crumbs coagulate into large clods then the soil cultivation is more difficult and plant roots cannot penetrate effectively as they are either waterlogged or suffer from excessive drought. Soils that contain very fine particles, with little crumb structure, easily form impermeable caps which dry into a solid mat, preventing penetration of shoots from emerging seedlings and resultant absorption of water. Consequently, seed germination is poor and erratic and subsequent crops fail to yield well.

Soil texture

The texture of soil describes the relative proportions of clay, sand and silt present and hence the physical components of soil crumbs. All three are basically composed of silica with possibly other elements such as aluminium. Sandy soils have the largest particles composed of sharply structured silica rock (see Figure 2.45). They hold very little water and are extremely free-draining and hence prone to drought.

Clay particles are composed of flattened structures with a high water-holding capability because they can store it on both the inside and the outside of each particle. They contain large quantities of other minerals and hence are a good source of nutrients for plants (see Figure 2.46).

Figure 2.45 *Sandy soil*

Figure 2.46 *Clay soil*

Figure 2.47 *Silt soil*

Unfortunately, because they attract and hold so much water clay soils can easily become waterlogged and are cultivated at times with difficulty.

Silt particles are much finer still and contain large amounts of nutrients particularly, because they have derivatives of older rocks known as feldspars in their structure (see Figure 2.47).

Ideally a fertile soil should contain all three components forming a healthy loam soil. The proportions of sand, silt and clay in loam are in the ratio 2:2:1 with about 5 per cent well-decayed organic material. The texture of a soil is usually determined by a finger test (see Figure 2.56).

Soil composition and plant growth

Soil composition influences their productivity and usefulness. Sandy soils are "early" and "light" soils because they absorb heat more quickly because of their low water content and require less energy invested in their preparation. Consequently, they are useful for growing crops out of season. Their disadvantage is proneness towards drying out and hence causing wilting by plants which then require remedial irrigation. Clay soils are capable of holding water and hence can continue supporting plants in dry periods. But their strong water retaining properties mean that during very dry weather plants cannot extract water from clay soils although it is present. Additionally, clay soils may become waterlogged in wet weather which makes cultivation and harvesting difficult. Clay soils are considered "heavy", demanding large investments in muscular or mechanical energy during cultivation. Soils were originally defined as "light" or "heavy" depending on the numbers of horses required for ploughing them.

If clay soils dry out for lack of rain or irrigation, then crack open up damaging crops. Cracked soils are very difficult to restore. Silt soils have similar characteristics to clays. They can be very suitable for vegetable crops particularly, because of their high nutrient mineral content. But they have a dangerous capacity for capping because of their fine particle size. Capping seals over their surfaces with an almost impenetrable layer and is caused by excessive rain or drought. Capped soils are difficult to cultivate. Continual good husbandry involving the addition of manures and composts will over time significantly improve soil health and quality by encouraging the activities of beneficial microbes and animals such as earthworms. Increasing health and quality raises the productivity of soils and reduces reliance on the use of fertilisers and watering.

Organic soils

Soils containing high concentrations of organic matter are usually derived from very wet marshland. Over many centuries, plant material decays, forming a deep layer of peat, which usually overlays a clay soil (see Figure 2.48).

Below both is rock. Organic soils can be extremely productive, providing good growing conditions for vegetables because they are rich in nutrients and water. Unfortunately, as the peat is cultivated it oxidises and as a consequence disappears over time. Cultivating peat soils bringing some of the underlying clay materials towards the surface adds to their fertility. Because of the high organic matter content, there are from time to time difficulties in using some weed-killers (herbicides), which are adsorbed onto the peat and hence their effectiveness is poor.

2.2 What soils provide

Soil is the source of nutrients, water, anchorage and biological support for plants (see Chapter 1). Soil provides plants with sources of nutrients and water from the time when seeds germinate or transplants are planted. These essentials are taken into plants through the roots and transported upwards in the vascular system and distributed into stems, leaves, flowers and fruit. Balanced supplies of nutrients are essential for healthy growth and reproduction. As discussed in Underpinning knowledge 2.1, soils are derived from rocks which contain mineral elements and then these are made available by weathering for uptake by plant roots. These supplies are supplemented by manures and fertilisers.

Figure 2.48 *Peat soil*

Nutrients are added as fertilisers into the soil increasing the yields and quality of vegetables and fruit and the beauty of flowers. Fertilisers fulfil two roles, improving the structure of soils increasing their biological health and quality and providing supplies of the individual nutrients taken in through the roots and used for constructing carbohydrates, proteins and lipids which underpin the growth and reproduction of all plants. Some fertilisers can be added as foliar feeds sprayed onto the aerial parts of plants, in all probability a considerable proportion of the nutrients applied to foliage is taken up through root systems from run-off (see Underpinning knowledge 4.4).

Improving soil structure and the micro- and macro-biological health and quality of soils is achieved by increasing their organic-matter content. Traditionally, this has resulted from adding farmyard manure during primary soil cultivation. Farmyard manure is available in rural areas but is less so in urban and sub-urban zones. The most effective form is horse manure, which generally is a balanced mixture of faeces and straw. It should be stacked for several months and preferably "turned", encouraging decomposition. "Turning" manure or compost heaps is the process by which the material is stirred, aerated and mixed, either manually or mechanically, and ensures that microbial decomposition is uniform and effective. Good-quality, well-turned horse manure has very little odour and appears as a dark-brown amorphous compost. Manures produced from other animals contains larger proportions of ammoniacal nitrogen which can encourage lush growth of plants which is often susceptible to diseases. Poultry manure, in particular, contains very large proportions of nitrogen unless well decayed and can damage plants.

Types of nutrient

For healthy growth, plants require supplies of four macro-nutrients. These are nitrogen (N), phosphorus (P), potassium (K) and calcium (Ca), which are needed in larger quantities compared with a range of micro-nutrients used in lesser amounts (the meaning of chemical symbols is given in the Glossary). Nonetheless, both macro- and micro-nutrients are essential and where one or more is in short supply plants show deficiency symptoms and do not thrive. The composition of plants is as follows:

- Green plants consist of 70 to 90 per cent water and 10 to 30 per cent dry matter.
- The dry matter consists of 90 per cent organic and 10 per cent inorganic material.
- Organic material is composed of carbohydrates, fats and proteins.
- Carbohydrates and fats contain carbon, hydrogen and oxygen.
- Proteins contain carbon, hydrogen, oxygen, nitrogen, phosphorus and sulphur.
- Inorganic material contains boron, calcium, copper, iron, magnesium, manganese, molybdenum, phosphorus, potassium, sulphur and zinc.

Nitrogen is the one nutrient which does not come from rocks to any large extent. It is derived by the actions of microbes in soil acting on gaseous nitrogen captured from the atmosphere and the effects of electrical storms (see Underpinning knowledge 2.3). The development of means for artificially preparing nitrogenous fertilisers from atmospheric nitrogen gas was one of the major triumphs of the early 20th-century industrial research. Nitrogen is particularly important for the growth of aerial organs, leaves and stems. Shortages of nitrogen result in poor growth, yellowing of leaves and is often accompanied by a bluish tinge in the lettuce and brassicas (see Figure 2.49).

Figure 2.49 *An example of nutrient deficiency: nitrogen deficiency in lettuce*

Figure 2.50 *Lime-induced chlorosis in* Wisteria *spp. grown against a wall*

Excess nitrogen causes rapid, soft growth which is increasingly susceptible to pests and diseases. Phosphorus is particularly important for effective root growth and encourages the establishment of seedling or transplant roots. Deficiencies result in poor root development with consequent stunting, yellowing and shrivelling of leaves, particularly in growing shoots. In the early stages of deficiency there may be speckling of leaves. Within plant cells phosphorus plays a vital role in the transport of energy during photosynthesis and respiration (see Underpinning knowledge 1.1 and 1.2). Potassium contributes to healthy fruiting and flowering and also enhances resistance to environmental stresses caused by cold, heat or drought. Shortages result in browning and shrivelling of leaves and failures in the development of flowers and fruit.

The fourth major nutrient is calcium, which has important roles in strengthening the walls of plant cells. It encourages activity at the sites of rapid cell division and subsequent growth at the tips of roots and shoots (see Underpinning knowledge 3.1). Calcium is also important as a constituent of chemically driven reactions within cells. Absence of this element is frequently seen as the die-back of leaf tips and internal browning. Soils derived from chalk or limestone rocks are usually well supplied with calcium. Organic soils and those derived from acidic rocks such as granite require regular additions of calcium as lime.

About 20 micro-nutrients are used in plant growth and functioning. Those which most frequently require supplementation by the gardener include: boron, copper, iron, magnesium, manganese, molybdenum, sulphur and zinc. Mostly these elements act as partners in chemical reactions which are driven by enzymes in cells. Their absence inhibits or reduces the efficiency of these reactions resulting in symptoms such as stunting, yellowing, shrivelling, blotching, spotting and eventual death. A number of other elements, such as sodium or selenium, are needed by plants, but usually in such tiny quantities that soils contain more than adequate supplies. Deficiencies which gardeners are most likely to encounter are iron and magnesium. This happens most frequently on soils which are derived from chalk and limestone. "Lime-induced chlorosis" appears where the plants turn yellow and fail because excessive calcium inhibits the uptake of other elements (see Figure 2.50).

Soil water

Soils provide plants with supplies of water. This is essential because all cell reactions take place in solution so that where water is in short supply plants rapidly wilt, shrivel and die from drought. Water is also important because nutrients are taken from the soil in solution entering through the walls of delicate single-celled root hairs which grow very rapidly at the tips of major roots. But excessive amounts of water can also be damaging because it drives out air, especially oxygen, from between the soil crumbs (see Figure 2.51).

Oxygen is essential for root respiration, and consequently soil waterlogging drowns the roots, ultimately killing the aerial organs. The aim of good gardening is to develop soils that hold sufficient water for satisfactory growth of the plants but which are also free-draining so that they allow an interchange of air from the atmosphere, providing ample oxygen and removing carbon dioxide. A soil that is effective in these respects will have pores between the soil crumbs holding equal quantities of water and air.

When soils receive rain, or are irrigated, they fill with water to their maximum holding ability, which is known as field capacity. When they reach field capacity, there is balance between water draining out under the influence of gravity and that filling the pore spaces and leaving sufficient openings for air which satisfies the respiratory requirements of plant roots. This balance is achieved relatively quickly in sandy soils which drain rapidly, holding relatively little moisture compared with clay soils where the water is retained strongly between the platelets (see Underpinning knowledge 2.1).

In addition to water removed from soils by plant roots there are also losses caused by evaporation from the soil surface. This process is accelerated on warm sunny days when the soil heats up and loses moisture or on windy days when soils are dried by forced evaporation. The gardener must be alert for either of these events as plants, especially seedlings and transplants, may rapidly become short of soil moisture and suffer from wilting or even death. Plants wilt when the aerial parts are losing more moisture to the atmosphere than is compensated by uptake from the soil. This condition may be reversible when irrigation is applied quickly or there is a spell of rain. Without the addition of moisture at some point plants will pass beyond the stage where wilting is reversible and it becomes permanent, in which case death follows quite quickly. Plants will suffer from wilting more quickly in sandy than in clay or silt soils because the latter are able to hold more water between their pores and attached to the soil particles themselves.

Figure 2.51 *Waterlogged soil*

Soil conductivity

The excessive addition of nutrients in soils can raise their conductivity. This is a chemical condition which in high concentrations

Figure 2.52 *Excessive salts deposited on organic soil*

of chemical ions will damage roots by scorching them (the meaning of "ions" is given in the Glossary). In a manner similar to the pH scale, excessive conductivity in soils is measured by the pC scale which is based on a measurement of electrical resistance. Measuring pC requires laboratory equipment, the results of tests are then converted into an arbitrary scale of 0 to 6. Crops such as lettuce and courgettes are susceptible to root burn from pC values greater than 3. Excessive pC values result in a white powder appearing on the soil surface (see Figure 2.52).

The simplest remedy for excessive pC values in soil is digging the ground and then allowing it to rest in a fallow condition for 12 months. During that time, rain will, most probably, have washed the excessive chemicals out of the land. Irrigation can be substituted for rain but large volumes of water are required, and if these are applied over a relatively short time the structure of surface layers of the soil can be damaged. Thereafter, in soils prone to salt accumulation avoid using large quantities of fertilisers, especially those prepared as liquid formulations.

2.3 The life of soil

Soil is a living entity filled with arrays of micro- and macro-organisms. Micro-organisms are those which are only visible with the aid of a microscope while macro-organisms are visible to the naked eye. Micro-organisms are bacteria and fungi; macro-organisms are a wide range of life forms from invertebrates such as earthworms, insects, slugs and snails through to vertebrates such as moles, mice, foxes, rabbits and badgers. The quality of a living soil is judged by its health, this is a measure of its biological activity and the way in which it benefits growing plants. The gardener aims always at increasing soil health. Initially when taking over a garden whether it is a new build or an established property or allotment, the gardener may find regrettably that the soil is in poor condition. This is typified by a soil which is compacted and easily becomes waterlogged. The gardener's aim will be to open up this soil, increasing its aeration and the percolation of water. Adding organic material either as well-rotted compost or farmyard manure into the soil is an essential first step towards developing better soil health (as described in digging soil). The gardener should view this as a long-term practice which should take place during each fallow season in late autumn and early winter. Building up the organic-matter content in soils on a continuing basis, over years, makes the growing of crops easier, more productive and of higher quality.

Soil microbes

Microbes are numerically and biologically the predominant inhabitants of soils. Most are microscopically small such that a gram of soil can contain millions of microbes, particularly bacteria.

Figure 2.53 *Nitrifying bacteria in nodules on broad bean roots*

These together with fungi are of critical importance and benefit. Their main function for the gardener is breaking down organic matter into its constituent parts. This process of degradation releases nutrients, growth regulators and stimulants, which can be taken up by plant roots and benefits growth. The breakdown of leaves, compost and manure is brought about by enzymes which both bacteria and fungi produce as exudates; exudates are solutions which pass out of the cells of plants or microbes into the soil.

In these processes, the bacteria use some of the resultant products as sources of energy for their own metabolism and secondarily by promoting plant root growth they benefit from their exudates. Roots release sugars and other chemicals which are used by the bacteria and fungi. Consequently, there is a cycle of reciprocal benefit which accrues for each of the partners in soil and is termed symbiosis. This is most evident with legume crops like peas and beans where beneficial bacteria form partnerships with the roots. Evidence of these partnerships is seen as white nodules on the roots (see Figure 2.53).

Bacteria in these nodules are capable of fixing atmospheric nitrogen gas into forms that can be used by the roots. In return, the bacteria gain sugars as energy sources from the process of photosynthesis provided by the plant.

The number of types of microbe in soil, particularly bacteria, is huge, and as yet only about 5 per cent have been identified scientifically and their capabilities described. This is one reason why it is very important that microbes involved in producing productive soils are cherished and preserved because there is so little knowledge of their beneficial properties and details of their lives. Bacteria are generally encouraged by soils with alkaline pH values, those greater than pH 7.0, while the fungi dominate mostly in acidic soils.

Mycorrhizal fungi

The benefits accruing from some fungi have been described scientifically in more detail. These are the mycorrhizal types which live in very intimate association with roots. They either form a sheath around the outside of roots as ecto-mycorrhizae or they enter the outer root cells and are known as endo-mycorrhizae These fungi gain greatly from sugars in the roots which have been produced by photosynthesis in the leaves. In return they are able to solubilise phosphorus compounds in the soil, which the roots then utilise. Recent research has also shown that mycorrhizae provide roots with protection from some pathogenic root-invading fungi. Hence, they enhance the health of the roots. Commercial preparations of mycorrhizae can be obtained from garden centres and mail-order suppliers. Adding mycorrhizae into the planting holes particularly for trees and shrubs, helps increase their root activity, growth rate and potential for survival. More widely, science is now showing that there are very close relationships between plants and soil microbes such that they are able to exchange information. Bacteria in particular respond collectively to changing environments in the soil, producing an array of chemicals with which they can challenge potentially damaging and aggressive pathogenic bacteria and fungi. The sophistication of information exchange between microbes and with and between plants is being revealed by studies in molecular biology.

Pathogenic microbes

Some forms of bacteria and fungi are less welcome because they cause plant diseases. Healthy soil has a balance between benign microbes and the pathogenic disease-causing types. When this balance is disturbed by cultivation, the pathogenic types begin to dominate and hence cause diseases. Soil-borne pathogenic bacteria and fungi cause forms of root rotting, collar and stem rot and some types penetrate into the root tissues causing massive distortion and loss of function and subsequent wilting. Initially, the infected crop plants will appear unhealthy and eventually they will die. Improving soil health is the gardener's first line of defence against disease-causing microbes.

It is particularly important that the gardener rotates vegetable and other crops. Rotation means that specific groups of vegetables should be grown only once in every four years on a piece of ground (see Table 2.1). Each component group in a garden rotation should consist of: potatoes, brassicas, legumes, onions and other alliums. Crops such as lettuce, courgettes and sweet corn can be grown as second crops in each section of the rotation after the main crop

Table 2.1 *An example of vegetable crop rotation*

Year 1	Year 2	Year 3	Year 4
Potatoes	Legumes – peas and beans	Alliums – leeks, onion, celery	Brassicas – cabbage, cauliflower, Brussels sprouts, broccoli, swede, turnip
Legumes – peas and beans	Brassicas – cabbage, cauliflower, Brussels sprouts, broccoli, swede, turnip	Potatoes	Alliums – leeks, onion, celery
Brassicas – cabbage, cauliflower, Brussels sprouts, broccoli, swede, turnip	Alliums – leeks, onion, celery	Legumes – peas and beans	Potatoes
Alliums – leeks, onion, celery	Potatoes	Brassicas – cabbage, cauliflower, Brussels sprouts, broccoli, swede, turnip	Legumes – peas and beans

Note: lettuce, sweet corn, courgettes may follow any of these groups has been harvested. In three of the four years brassicas follow the legumes. This maximises the value derived from the residual nitrogen left from the nitrogen-fixing bacteria which inhabit legume roots. Rotations also help prevent a build-up of disease-causing microbes in the soil as, for example, onion neck rot (*Botrytis allii*) (see Figure 2.54) or clubroot of brassicas (*Plasmodiophora brassicae*) (see Figure 2.55) which results from the accumulation of the disease-causing microbe in the soil. Crop rotations were developed during the early 18th century as the "Norfolk husbandry" which introduced legumes and turnips in rotation with cereals, which boosted yields and obviated the need for fallowing.

In the time periods between each section of the rotation beneficial microbes will erode the populations of the disease-causing types. In these bio-control processes, the gardener is making use of the warfare between benign and pathogenic microbes, encouraging the suppression of the latter (bio-control or biological control is the use of a beneficial macro- or micro-organism as a means of combating pests, weeds and diseases, see also Underpinning knowledge 4.3). In encouraging such suppressive action, the gardener should avoid the excessive use of nitrogenous fertilisers particularly those containing ammonia-based compounds. In this respect, the use of calcium nitrate and calcium cyanamide fertilisers is especially valuable. These fertilisers stimulate the activity of benign microbes which destroy the plant disease causing species. The gardener may also guard against crop pathogens by using resistant varieties of vegetable and fruit where these are available. Resistance is the simplest, most efficient and easily applied form of disease control because it comes built into a crop by the plant breeder and it operates as soon as the seed is sown.

Ultimately, however, the incidence of diseases on crops may become so frequent that the gardener must resort to using chemical controls. Mostly, these are applicable for diseases of foliage and fruit and so do not interfere with the natural balance of health in the soil. Very few, if any, chemicals that will combat soil-borne pathogens are now available for the gardener. Ideally,

Figure 2.54 *Onion bulb rot caused by the fungus* Botrytis allii *which is seed-, bulb- and soil-borne and may also be imported in diseased onion sets*

Figure 2.55 *Clubroot of brassicas caused by the protist microbe* Plasmodiophora brassicae

chemical controls should be applied as soon as initial symptoms start appearing. At this stage, chemical control is at its most effective. Persistent diseases will require several applications of control chemicals.

Soil macro-organisms

Substantial numbers of macro-organisms, those visible to the naked eye, also inhabit soil. Probably the key species in soil are earthworms which Charles Darwin christened "Nature's plough" as he recognised their importance in developing soil health. The gardener will find that as increasing quantities of organic matter are added to the soil over years so populations of worms rise. This is a major sign that soil health is improving as worms take in and digest soil through their intestines. They excrete residues as worm castes, which are very rich in mineral nutrients and hence increase fertility. Burrowing by worms opens up drainage and aeration channels in the soil so improving water and air percolation.

Soils also contain other invertebrates. Some are beneficial such as types of springtails (*Collembola*) and nematodes (*Nematoda*) which destroy other plant-pathogenic types. Beneficial nematodes in particular will destroy other pest members of their group by capturing them by snaring. Wireworms (a form of beetle belonging to the family *Elateridae*) are very destructive since they will burrow into the roots of crops and especially potato tubers rendering them un-useable. Cabbage (*Delia radicum*) root flies and cabbage white butterflies (see Figure 4.65) lay eggs near the collars of cabbages or on leaves, respectively, which subsequently develop into larvae feeding on the roots, channelling and ultimately destroying them or consuming the leaves. Some aphids, for example the lettuce root aphid (*Pemphigus bursarius*), also inhabit soil and are difficult to destroy once an infection has started. Each of these pests can be deterred but not wholly eliminated by using crop rotations, preventing a build-up of larger populations. Some nematodes and a few fungi transmit virus diseases into crops. These cause losses to the vigour of plants and are usually first noticed because of yellowing, mottling and scorching of foliage. There is little that the gardener can do that will directly eliminate these pathogens except for using crop rotation and regular cultivation and incorporation of organic manures.

Modern housing developments are restricting garden size which limits opportunities for using rotation as a means of pest and disease control. This means that gardeners may limit the frequency with which particularly vulnerable crops such as brassicas are grown. Use may also be made of companion plants such as varieties of pot marigold (*Calendula officinalis*), herbaceous plants which release chemicals from their roots and are antagonistic towards some disease-causing microbes.

A few vertebrates particularly mammals inhabit soils. Field mice (*Apodemus sylvaticus*) shelter in burrows, possibly those formed by other soil inhabitants. These pests can cause substantial damage especially in root crops like carrots and beetroot and will eat peas from the vine. Mice are also particularly partial to the emerging shoots of species gladioli in the early spring. Voles are also a scourge for bulb gardeners since they will destroy plantings of corms, for example crocus is a favourite target for foods. Their predation can be avoided by planting crocus in particular "in the green", which allows the plants to become established in late spring before these animals are hunting for foods in late autumn and winter. Mostly these animals are seeking roots, seeds, bulbs and corms, which contain large quantities of sugars. Trapping can keep these animals at bay but ultimately their source requires identification and destroyed and that is a task best left to professional vermin control officers. Gardens or allotments may be inhabited by larger mammals such as rats, squirrels, rabbits, foxes and badgers. Whether some of these are welcome or not is a matter of individual personal choice for the gardener and his or her family. They will cause, however, very considerable damage to crops and soil health, making some aspects of gardening impossible.

2.4 Dirty hands

Physically handling soils is the best way of understanding what constitutes their health and fertility. Don't be reticent about handling soil, it is after all the most basic gardening ingredient.

Soil quality

The simplest and quickest means of testing the quality of soil is the finger and thumb test. This indicates the texture of soil in the simplest terms. It indicates qualitatively the proportions of sand, clay and silt particles that are present. From that, the gardener will learn whether the soil is easily worked and warms quickly in spring and, therefore, encourages the speedy germination or rooting of transplants. This is the case with soil which contains large proportions of sand particles and is a "light soil". Alternatively, the soil may contain large amounts of clay or silt. Clay, "heavy" soils are water-retentive but slow to warm up and easily become waterlogged. Silt soils, while being very productive, especially of leafy vegetables, cap easily, making germination difficult.

Finger test

In this test, a small amount of soil is rolled between a wet forefinger and thumb (see Figure 2.56).

Soils with a preponderant sand content feel sharp because the particles have clear-cut edges which grind against the skin, clay soils will feel sticky and can be moulded quite easily into a sausage shape and silty soils have a silky feel because of the very small size of their particles. Obviously, most soils will have a proportion of each type of particle but with experience the gardener can judge these relative proportions. One way to gain such experience is to feel pure sand, clay (as used by potters) and silt from the side of a river or stream after it has flooded. Having experienced these pure constituents, the gardener can then progress to testing a variety of soil types.

Figure 2.56 *Soil finger testing*

Soil flotation test

A more visual test of the composition of soils can be obtained by a flotation test. This requires a large jug, jam jar or any container with clear sides, water, a short bamboo cane and a small sample of soil. Fill the container with water and place the sample, 50 to 100 g in 500 ml water, of soil on top of the water and stir with the cane. Leave the container on a solid surface for 24 to 48 hours. At the end of this time, gravity will have separated out the soil content according to the weight (mass) of the particles into silt, clay, sand and organic matter (see Figure 2.57).

Figure 2.57 *Soil flotation test*

Sand, being the heaviest, will be found in a layer at the bottom of the container; above that will be a layer of clay particles, and above that the silt. The relative thickness of these layers will indicate the proportions of their components. At the top of the container will float a layer of organic matter because this has a density which is less than that of water. The thickness of this layer will indicate whether the soil is organic derived from origins in peat bogs or is a mineral soil. The latter is likely for most soils. This test can be repeated with soil samples taken from round the garden in order to understand their variation in composition and potential for use in vegetable, fruit or ornamental cultivation. It is also worthwhile repeating this process annually especially for the vegetable garden which will be receiving regular applications of well-rotted compost or farmyard manure. Recording the findings from these tests over several seasons will establish how well the soil is responding in terms of developing a healthy profile. It is not easy to increase the content of sand, clay and silt, but the organic matter content should rise over years as more humus accumulates in the soil as a result of adding farmyard manure or compost. That will indicate that the soil is healthier and the crops should gain from a greater retention of soil moisture, which reduces the need for irrigation in warm, dry periods.

Soil profile

Digging a soil profile pit (see Underpinning knowledge 2.1, see Figure 2.44) is one of the simplest ways to understand whether the soil will become waterlogged after heavy rain or is very free-draining and likely to need irrigation during droughts. It also indicates especially for new-build gardens how much damage has been done to the soil by the builders and what they may have left behind in the soil. A soil pit is simply what it says. Using a digging spade, take out a hole that is 75 cm square, going down as far as possible. If the soil is very deep, for example deeper than 1.5 m before the rock is encountered, then it is probably wise to assume that the pit is deep enough. Few new gardens are likely to have soil depths greater than 2 to 3 m. The gardener should be able to identify the layer of rotting leaves and other organic material sitting on top of the soil. Beneath that the various horizons of top soil, subsoil and rocky material should be obvious (see Figure 2.44). Of most interest will be the identification of panning at some depth in the soil. Pans are layers of almost impenetrable hardened soil. This may be a natural pan or most likely in a "new-build" garden will be the result of heavy machinery having trafficked across the land during building operations causing consolidation. Pans prevent water from percolating through the soil and plant roots are frequently unable to penetrate through them. Consequently, they can lead to waterlogging and to unthrifty plant growth which is prone to failure in times of environmental stress such as periods of drought. If there is a pan in the garden, then the most effective remedy is to enlist the aid of a garden-landscaping company which can use a small tractor or rotavator fitted with a pan-busting tine which penetrates and disrupts it.

Soil auger tests

The gardener may also feel as he or she becomes more experienced that they wish to understand the soil structure of the entire garden. This can be achieved by use of a soil auger which will take cores from the ground and allow their examination. Small hand augers can be purchased from mail-order companies. They are made in two patterns, either a screw thread (see Figure 2.58) or a lance (see Figure 2.59).

The former will take deeper samples. The lance type is mainly used for sampling the top soil and is suitable for most gardens.

Samples should be taken in a "W" pattern across the garden. Those taken with the screw thread will offer a comprehensive view of the entire soil profile, and the gardener will be able to recognise the topsoil, subsoil and bottom horizons. From these, take finger test samples and explore the way in which soil texture changes

Figure 2.58 *Helical soil auger: commercial type*

Figure 2.59 *Lance type soil auger*

down the horizons. This gives a good view of the potential of the garden for adequate drainage and aeration. They will also warn of the prospect for waterlogging and impeded drainage. The gardener will also see how soil depth varies across the garden. Such variations can occur over relatively short distances. This gives a useful indication of where deeply rooting subjects such as fruit trees can safely be planted.

Soil pH tests

One important factor which gardeners need to establish is the acidity or alkalinity of the soil. This will regulate the types of ornamentals which can be grown and also indicate the likelihood of encountering problems with some vegetables such as tip burn in lettuce caused by high alkalinity and excess calcium.

Plants are of two types in relation to soil pH, either calcicole, which grow well in alkaline soils, or calcifuge, which require acidic conditions. Acidic pH soils are used for calcifuge (lime hating) plants which cannot tolerate chalk and limestone. This includes ornamentals such as *Rhododendron*, *Magnolia* and *Camellia* as well as a variety of rock garden plants. Alternatively, calcicole (lime-liking) plants require alkaline pH soils, including many of the cherries. Testing soil pH is relatively simple, kits are obtainable from the garden centres or mail-order suppliers. These consist of a set of instructions, several test chemicals and some small plastic vessels in which the tests are made. Take a small sample of the garden soil, place this in one of the dishes provided and add the chemicals as instructed. This will result in a colour change reaction as the chemicals are affected by the acidity or alkalinity of the soil. The resultant colour is matched against a chart which reads off the pH in a series of steps. Additionally, garden centres and mail-order companies can supply pH probes, which are simply inserted into the soil and will provide an immediate reading (see Figures 2.60 and 2.61). Some newer probes will also offer measurements of soil moisture and light levels.

The pH scale from 0 to 14, with 0 to 7 being acid and 7 to 14 being alkaline, has its basis in physical chemistry but its derivation need not concern gardeners. The extremes are beyond any

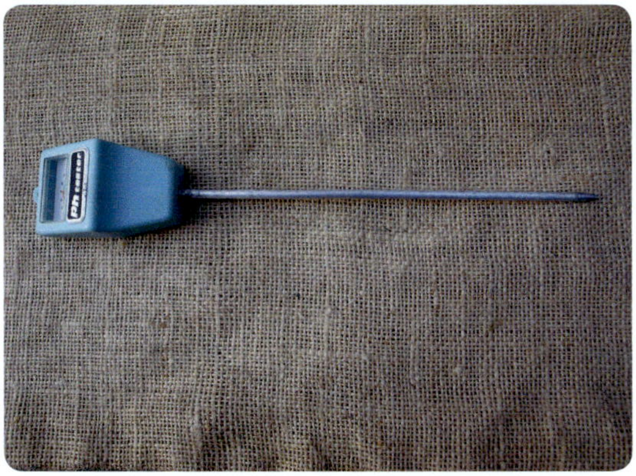

Figure 2.60 *Soil pH tester with probe used for assessing acidity and alkalinity*

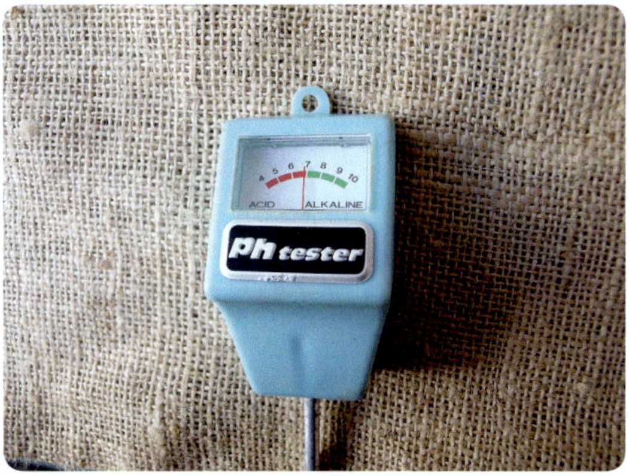

Figure 2.61 *Soil pH tester showing the scale between acidity and alkalinity*

value likely to be encountered in gardens. Generally, soils will have a pH in the range 6.5 to 7.5, slightly acid or slightly alkaline, which is what is required by most crops. If the pH is below 6.5, then the gardener should lime the soil. Lime is available from garden centres. Applications of 30 g per square metre will raise pH slowly by about 0.5 unit. This effect is not instantaneous so that lime is best applied after digging and is then washed into the soil by winter rains.

Soils where the underlying rock is chalk or limestone will not require liming. But testing the soil pH of the soil even in areas which are predominantly calcareous is an advisable precaution. Soil types may vary over short distances with pockets of acidic rock such as greensand present. Additionally, builders may "pollute" land with calcareous materials especially near houses, which can then give a false impression of the garden's soil type. Observation of surrounding gardens will indicate the most likely general soil pH value.

Lime is usually bought as crushed chalk or limestone rock which are delivered in varying degrees of fineness. These are forms of lime which will affect soil pH about six months after application. There are more rapidly acting forms of lime. Crushed limestone rock, calcium carbonate, when roasted at high temperatures, produces calcium oxide. This is a highly reactive substance that produces large amounts of heat when mixed with water. Calcium oxide is not suitable for gardens since application requires specialist equipment and knowledge. When calcium oxide reacts with water, it produces calcium hydroxide or slaked (hydrated) lime in a very fine powdered form. Slaked lime may be used in the garden and is sold in small quantities in garden centres and by some mail-order suppliers. Calcium may be added into soils without affecting pH by using crushed magnesian limestone. Calcium chloride is a rapidly acting and soluble form of lime used for the reduction of bitter pit and internal browning in fruit. It can be sprayed on to growing crops.

Calcium cyanamide and calcium nitrate are two forms of proprietary fertilisers delivering both calcium and nitrogen for plants. Calcium cyanamide is a slow-release form of nitrogen which affects soil pH very slowly and encourages populations of benign microbes which are antagonists of soil-borne pathogens. Calcium nitrate is useful particularly for leafy vegetables providing a rapid growth stimulus and reductions in soil borne diseases. Both these fertilisers supply nitrogen in a nitrate form which encourages steady growth. That is compared with the rapid bursts of growth associated with ammoniacal forms of nitrogen. Rapidly growing crops are more susceptible to pest and disease attacks.

If soils are in excess of pH 7.5, then they should be allowed to diminish alkalinity naturally by the effects of atmospheric carbonic and nitric acid in rain water. Reducing alkalinity by artificial means is a slow and not very reliable process. If the gardener wishes to do this, then it is best achieved by applying sulphur dust, which in soil moisture converts to weak sulphuric acid in the soil. Sulphur dust or "flowers of sulphur" can be obtained from some garden centres or mail-order suppliers. When applying either lime or sulphur dust, the gardener should wear a mask preventing small particles from being taken into the throat and lungs and protecting hands with gloves. Children and pets should be kept away from these processes. Once rain has fallen on these materials, then their dustiness will be safely abated.

Iron sulphate is particularly useful where plants are suffering from calcium-induced iron deficiency and the leaves are showing considerable yellowing (chlorosis). Excessive calcium in soils blocks the capacity of roots from absorbing other elements such as iron. Combating lime-induced deficiencies can also be achieved by using chelated, or fritted, forms of fertilisers. In these compounds, a particular form of chemical bonding for elements such as magnesium or iron holds the atoms very tightly until they are released near the plant roots. Magnesium or iron are then released and absorbed into the roots and translocated towards the leaves where deficiency symptoms are alleviated.

Straight and compound fertilisers

Fertilisers are available as 'straights' or 'compounds'. Compound fertilisers contain several nutrients. The composition of fertilisers is regulated by law because they are used in large quantities in agricultural and horticultural industries and there the users must be able to rely on the accuracy of labelling. The container bag will list the nutrient elements present in the fertiliser and indicate their quantities in percentage terms (see Figure 2.62).

Nitrogen (N) will be described in terms of ammonium nitrogen or nitrate nitrogen; phosphorus as phosphorus oxide (P_2O_5); and potassium as potassium oxide (K_2O). Lime likewise will report the amount of calcium oxide (CaO) in hydrated lime, and other mineral nutrients may also be quoted as oxides. For the major nutrients, the gardener requires a balanced fertiliser in terms of its N:P:K content, for example, a compound fertiliser such as "GrowMore" (see Preamble) which has a nutrient ratio of 1:1:1 or permutations (N:P:K). Where ample nitrogen is being derived from the application of composts and farmyard manure, then ratios of 0:1:2 (N:P:K) will be adequate. Other elements are also declared, such as boron (B), sulphur (S) or magnesium (Mg, or MgO indicating magnesium oxide). Compound fertilisers provide the basic dressing when seeds or plants are sown or planted in the garden. They are used as top dressing for established trees, shrubs, perennials and bulbs.

Straight fertilisers supply one or a few nutrient elements such as potassium nitrate (KNO_3) which supplies potassium and nitrogen. Single-element "straight" fertilisers are used either where there is a need for a boost to growth or particular deficiencies develop. They are also available as various sulphate salts such as potassium or magnesium sulphate. Potassium nitrate supplies both potash, which supports fruit and flower growth and an assimilable form of nitrate nitrogen. Fertilisers such as superphosphate, triple superphosphate or diammonium phosphate are used where there is deficiency in phosphate (P). Diammonium phosphate is a more soluble form and hence more readily available for plant roots. Phosphorus is a key element in plant metabolism, being of major importance for energy transfers

Figure 2.62 *Fertiliser analysis declared on the container (13:13:21, N:P:K)*

(see Chapter 1). While deficiencies will become very quickly evident from the rapid loss of growth, flower abortion and fruit drop, correcting them is a slower process.

Liquid fertiliser formulations are particularly valuable for feeding potted and glasshouse-grown plants. Formulations deliver the major nutrients in differing proportions dependent on the growth and fruiting stages of the plant. Their formulations are tailored either for encouraging fruit production or increased leaf formation depending on the ratios of nitrogen or potassium present. Large amounts of nitrogen stimulate leaf production whereas potassium encourages root and fruit growth.

Liquid fertiliser formulations may also contain micro-nutrients, which are frequently needed for potted plants. These types of fertiliser are useful in automated systems which feed and irrigate plants, a process known commercially as "fertigation". Solid forms of micro-nutrients are supplied normally as water-soluble preparations frequently as single element preparations delivering boron, copper, iron, magnesium, manganese, molybdenum, sulphur or zinc. Only in specialised circumstances are these likely to be required. It is unlikely that either macro- or micro-nutrient deficiencies will be encountered by gardeners, provided farmyard manure, composts, compound fertilisers and other general products are used as suggested, because they will contain very adequate quantities of the mineral elements required by plants.

Fertilisers are also available as "slow-release" formulations where the granules are coated with resins which dissolve at predetermined rates depending on the soil temperature. That provides plants with regulated quantities of nutrients over extended periods up to 12 months.

Gardeners will find several organic fertilisers on the market. In particular, these include dried blood, ground hoof and horn meal and poultry pellets which are becoming widely available as a source of nitrogen. The first two are useful as slow-release nitrogen fertilisers, which can be used especially for fruit trees. Seaweed extracts are available as concentrated liquid formulations. These have benefits for reducing the impact of soil-borne diseases, encouraging benign microbial populations in soil, reducing the impact of low temperatures and increasing the quality of fruit and vegetables.

Fertiliser mixtures termed compost teas are suitable for use in the organic culture of fruit and vegetables. These are available in garden centres or from mail-order suppliers. The gardener mixes the contents of a series of packages together in a container of water and allows the product to ferment before watering it around plants or spraying onto the foliage. Otherwise the gardener may make his/her own mixtures for organic husbandry such as that obtained by growing and then fermenting Russian comfrey (*Symphytum officinale*). The gardener can also prepare his own liquid fertilisers, albeit of unregulated composition, if there is a family pet rabbit. Retain some of the residues from cleaning the rabbit hutch and place these in a hessian sack. Suspend the sack in a container of water and leave for 1 or 2 days. During that time, the nutrients contained in the rabbit's residues will have leached into the water and may be diluted and used as a liquid fertiliser. Predominantly, the liquid will contain nitrogen, and there will be no simple means of accurately estimating its concentration. Consequently, the diluted solution should be used cautiously on well-established plants and not on young seedlings.

Green manuring crops are increasingly popular. Areas of the vegetable garden may be sown with rapidly growing plants such as Caliente mustard (*Sinapis* spp.) or Russian comfrey

(*Symphytum officinale*) in the autumn. These will produce a mass of foliage and roots which can be incorporated into the soil during the annual digging process. They rot in the soil as a result of bacterial action providing sources of nutrients and possibly some organic chemicals which are antagonistic towards pests and pathogens. This is a way of utilising natural processes as a means of reducing populations of crop-damaging organisms. This is a form of biological (bio-) control known as bio-fumigation. The effects are obtained by the release of *iso*-thiocyanates from the decaying mustard which is toxic predominantly to pathogenic microbes. Increasingly, home composting is being encouraged. Success requires that sufficiently high composting temperatures (in excess of 60 °C) are achieved which will destroy weed seeds, pests and microbes that cause plant diseases. The end product may then be useful as a soil improver.

Proprietary composts either in bagged form or in bulk loads can be very effective soil additives. Either dig them into the garden or use them as mulches laid over the soil surface, where they will decompose and increase soil fertility and deter weed growth. Composts can be obtained from municipal and commercial sources in bulk loads. Some arboricultural companies compost tree and shrub residues, which are useful mulches.

Professional soil sampling

A really thorough approach for developing "new" gardens is achieved by having samples of soil taken professionally and the resultant samples analysed in a commercial soil science laboratory. Analyses may be purchased for relatively modest outlays and are returned with professional recommendations. This approach will save the gardener money in the long-run by ensuring that fertiliser is used correctly, most efficiently and where it is needed most.

Achievements: have you understood and learnt?

At the end of the chapter, you should be able to:

1. Know how land is cultivated by digging, adding organic manures and prepared for planting crops of potatoes and onion bulbs.
2. Know how seed potato tubers and onion sets are cultured under protection before planting in the open.
3. Know the subsequent culture of these crops up to harvesting.
4. Understand that soil provides plants with rooting stability, nutrients, air, water and biological partners.
5. Know that soil is divided into topsoil, subsoil and degrading rock.
6. Understand that soil structure relates to the size and proportions of particles aggregated into crumbs which allow the free movement of water, air and plant roots.
7. Describe soil texture as the relative proportions of sand, clay and silt fractions and their properties.
8. Understand that soil particles aggregate into crumbs which hold air and water for use by plant roots.

9. Know that plants take up nutrients dissolved in water held in soils and that plants need oxygen from air in the soil in order to respire and release energy used in the cells.
10. Know that healthy plant growth requires four macro-nutrients: nitrogen (N), phosphorus (P), potassium (K) and calcium (Ca), plus a range of micro-nutrients which are used in lesser quantities but are still essential for growth.
11. Know that shortages of macro- and micro-nutrients cause mineral deficiency disorders in plants which reduce productivity and eventually kill the plants.
12. Describe that soils vary in acidity and alkalinity values, which is measured by their pH on a scale of 0 to 7 and that soils of pH 6.8 to 7.4 are most satisfactory for plant growth.
13. Know that adding lime will increase the alkalinity of soils over an extended period of time.
14. Describe that soils are living entities and are judged by their health and that soil health is a gauge of their biological activity.
15. Know that increasing the organic-matter content of soils builds up the populations of soil beneficial microbes.
16. Understand that microbes are key species in helping break down leaves and other organic matter into humus and releasing and recycling nutrient in the soil and building up the organic matter content.
17. Describe that beneficial microbes combat pathogenic forms and so help plants survive attacks by plant pathogens.
18. Understand that a range of macro-organisms also inhabit the soil, these are both invertebrates and vertebrate types and that worms are probably the most valuable invertebrate being called "nature's plough" by Charles Darwin.
19. Know that the texture of soils can be determined by the finger test which identifies the relative proportions of sand, clay and silt particles and that the proportions of organic matter, sand, clay and silt may be determined by a flotation test.
20. Understand that gardeners should become conversant with the manner by which the composition of fertilisers is expressed and the usefulness of different types.

Definition of terms: *Describe* is able to show why a scientific principle is important; *Understand* applies scientific principles into the context of gardening; *Know* carries through scientific principles in practical gardening activities.

Chapter 3
Growing legumes from seed and seedlings

Introduction

Seeds are used either by directly sowing into the garden or by raising transplants under some form of protection and then planting them out (transplanting). This chapter explains both forms of husbandry and the science underpinning the availability of healthy and viable seed which is true to type. The term "husbandry" has its origins in Middle English (AD 1150 to 1500) describing the careful culture of crops.

Garden practices

a. Seed origins and care

Plant breeding and the subsequent production of commercial quantities of seed is an expensive, sophisticated and extended process. It takes 15 to 20 years for plant breeders to identify and insert the genetic characteristics which will improve plant varieties and for these to be tested and accredited under commercial conditions. All agricultural and vegetable seed which is sold is carefully regulated by law, ensuring that it is pure, reliable and capable of producing high-quality crops as claimed by the producer. This is particularly important as seed is an internationally traded commodity and international treaties regulate the production, analysis and quality of seed of food crops and to a lesser degree flower crops. Consequently, international agreements between governments are required since in many instances seed is the basis for worldwide food supplies.

Gardeners obtain their seed either from a garden centre or by mail order from suppliers. Seed catalogues offer a bewildering array of varieties of every conceivable vegetable (see Figure 3.1). Reading catalogues and selecting seed for the next season can be an enjoyable winter pastime.

Figure 3.1 *Seed suppliers' catalogues*

The new gardener can be guided by descriptions in the catalogues, reports in the gardening press and by information provided by organisations such as the Royal Horticultural Society (RHS). Alternatively, gardening or allotment clubs can be good sources of local guidance from members who have been growing fruit, flowers and vegetables in the area for some time. Local experience and advice can be very helpful because varieties which thrive in one locality may fail in another neighbourhood simply because of the effects of minor variations in soil types or slight climatic differences.

The gardener is well advised to buy new, fresh seed each year. Attempts at saving seed produced from a crop of runner beans, for example, at the end of the season is a false economy. Self-saved seed will quite probably be infested with pests or microbes such as bacteria, fungi or viruses. If the variety in question is a hybrid, then the self-saved seed will not reproduce the same variety in the following season. This is because hybrids must be reproduced for each new seed batch by remaking the crosses between the original parental lines (see Underpinning knowledge 3.1 and 3.2). These breeding lines are carefully protected and safeguarded by each seed company.

Once seed is bought from whatever source, it should be stored in a cool, darkened place in a tin or tub with a tight lid (see Figures 3.2 and 3.3).

On no account should seed be exposed to strong sunlight or heat. Seed is a living entity, albeit in a dormant state, and its viability is easily harmed by poor handling and storage.

Figure 3.2 *Seed packets displayed in a garden centre: vegetables*

When required for sowing, seed is carefully removed from its package. It is quite possible that seed packets will be double-sealed. There will be an outer envelope displaying the name and some of the qualities of the seed and also a statement indicating the extent of its longevity and purity. Within that is a foil package which contains the seed. Foil packaging helps preserve the seed by preventing desiccation and giving some insulation from fluctuating temperatures, thereby safeguarding its shelf life. The atmosphere inside the foil packet will have a low relative humidity which is purposely designed so that seed is kept in a favourable state unaffected by microbial rots.

Figure 3.3 *Seed packets displayed in a garden centre: flowers*

The basic requirements needed for seed germination and plant growth in terms of warmth, water, aeration with oxygen, nutrient supplies and anchorage are described in Chapter 1.

b. Growing peas and beans

Legumes, peas and beans, are valuable garden crops and also benefit soil fertility. They can be eaten either fresh or blanched and deep frozen for future use. The roots of legumes form associations with soil-inhabiting bacteria, which are capable of fixing atmospheric nitrogen into forms that green plants can utilise (see Chapter 2).

The white nodules on pea and bean roots are manifestations of the activities of the beneficial nitrogen-fixing bacteria (see Figure 2.53). These bacteria are naturally present in soils in countries such as Great Britain because of centuries of growing legumes. As the pea or bean seed germinates, these bacteria are stimulated into activity, entering and colonising the roots, and they begin fixing atmospheric nitrogen. The plant gains access to nitrogen and in return the bacteria utilise carbohydrates manufactured by photosynthesis (see Underpinning knowledge 1.1) and transported to the roots and into the bacterial nodules. This is another example of the two-way process of mutual advantages known as a symbiotic relationship (symbiosis is a close relationship between two organisms which is advantageous for both). All legumes host forms of symbiotic natural associations with nitrogen-fixing bacteria consequently, they are valuable parts of crop rotations in the garden.

c. Germinating and growing broad bean and pea seed as transplants

Producing pea and broad bean seedlings under protection for subsequent transplanting gives the gardener greater flexibility and control in timing these crops.

Use a very large seed tray and line the bottom with some old newspaper and fill with seed compost to within 2 cm of the top (see Figures 3.4 and 3.5). If large trays are not available, then modular trays are equally appropriate.

Figure 3.4 *Long seed tray*

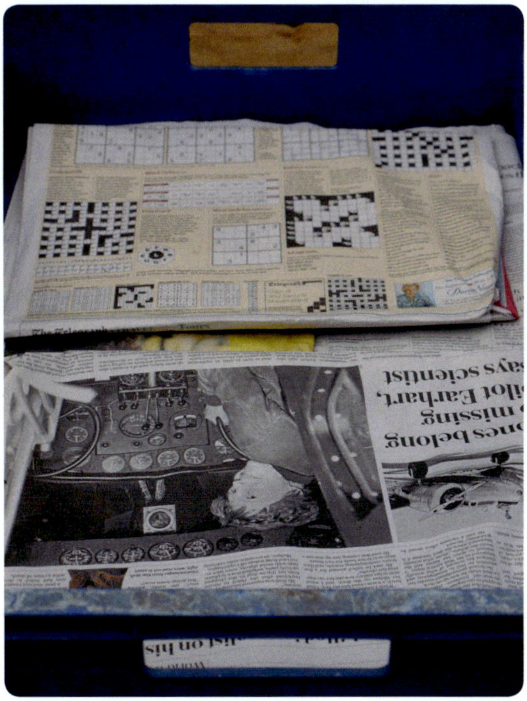

Figure 3.5 *Long seed tray lined with old newspaper*

Lightly press the compost down so that it settles, ensuring there are no large lumps left (see Figure 3.6).

This is especially important if the compost is taken from a compressed bale. Composts are heavily compressed in the production factories so that the bags are of minimum size commensurate with their stated volume.

Carefully open the seed packet and place the broad beans into a plastic saucer so that they can be picked up individually. Normally there is a greater quantity of pea seed and it may be sown directly from a packet. Take out each bean seed and place in rows across the tray, nestling each seed gently into the compost surface with about 3 cm between seeds within and between the rows (see Figures 3.7, 3.8, 3.9 and 3.10).

Figure 3.6 *Long seed tray filled with compost and firmed with a smaller tray*

Figure 3.7 *Spaced-out broad bean seeds on compost*

Figure 3.8 *Close-up of broad bean seed spacing*

Figure 3.9 *Garden sieve with broad bean seed*

Figure 3.10 *Garden sieve filled with compost*

Figure 3.12 *Cover with glass and sheets of old newspaper*

Figure 3.11 *Cover with compost and water well*

Figure 3.13 *Pea seed*

Place the tray(s) of broad beans and pea seed on the greenhouse bench and water with a fine rose fitted to the watering can and facing upwards, so that the water falls gently onto the compost (see Figures 3.11 and 3.12).

Pea seeds should be sprinkled from the saucer over the surface of the compost, ensuring that they are evenly spaced, occupying the entire box right towards the edges and into the corners (see Figures 3.13 and 3.14).

When sowing is completed, sprinkle more compost through the mesh of a garden

Figure 3.14 *Pea seed sown onto compost in a long tray*

sieve onto the seeds until they are covered to a depth of 2 cm. If seeds are uncovered by watering, then recover them with more sieved compost.

Place a sheet of glass over each tray and some sheets of old newspaper on top. This keeps the tray warmer and retains moisture. Trays should be inspected daily and the glass reversed, removing accumulated moisture. Once the shoots start appearing, remove the newspaper. The glass pane may be left in place for a few more days retaining moisture, but turning it over at daily intervals. This ensures that the air supply to the tray is replenished and there is no stale moist air above the developing seedling which predisposes them to diseases. Remove the glass before the seedlings reach its surface, again preventing damage and possible disease development.

d. Transplanting seedlings

Once the seedlings are 5 to 10 cm high, they can be planted out in the garden (see Figure 3.15). This will probably be about 15 to 20 days after the seedling leaves begin emerging through the compost. Since these transplants have been protected, gardeners should ensure that they are not planted out when the weather is adverse such as with forecasts of heavy rainfall or night frosts. Prepare the planting area carefully (see Figure 3.16).

Place the walking board on the surface of the land adjacent to the position where the peas and broad beans are to be planted. Rake the soil over vigorously, producing a fine tilth and sprinkle on some "GrowMore"-type general fertiliser at 30 g per square metre. Rake in the fertiliser and establish a straight row with a garden line as described in Chapter 2 (see Figures 3.17 and 3.18).

Gently tap one end of the seed tray on the ground to loosen the seedlings and then carefully tease them out from the tray (see Figures 3.19, 3.20 and 3.21).

Take out one broad bean plant at a time and do not allow the roots to dry out during transplanting. Pea seedlings may be planted in small bunches containing 3 to 5 plants. Make holes 10 to 15 cm apart with a garden trowel and place a broad bean plant or group of pea seedlings into each hole. Draw soil over and round the root systems and firm into the ground with gentle

Figure 3.15 *Pea seedlings ready for transplanting*

Figure 3.16 *Clearing the ground of germinating seedling weeds prior to transplanting pea seedlings*

Figure 3.17 *Rake the soil down to a fine tilth*

Figure 3.18 *Rake in a dressing of granular fertiliser*

Figure 3.19 *Standing the seed tray on its edge to loosen seedlings*

pressure from your fingers. As soon as transplanting is completed, water well, even if the soil is moist (see Figures 3.22 to 3.28).

Figure 3.20 *Planting pea seedlings with a trowel*

Figure 3.21 *A planted group of pea seedlings*

Figure 3.22 *Rows of planted bunches of pea seedlings*

Figure 3.23 *Watering in the transplanted pea seedlings*

This settles the soil around the roots. It is essential that the roots of the seedlings are able to have immediate access to water and nutrients from the soil so that they can continue providing them for the leaves.

Transplants are very vulnerable to attack from pigeons and slugs, so it is advisable to cover with a plastic or glass cloche or pieces of wire netting cut to size and rounded into a tunnel shape placed over the plants (see Figure 3.24).

Slugs should be controlled either using a form of biological control which is available from garden centres or via purchases made on the world-wide-web or chemical pellets.

Once the plants are established, they will require to be supported. For broad beans, this is best done by placing bamboos canes, 1.0 to 1.5 metres high, at 30 to 40 cm distances around the row(s). Tie garden twine (4 to 5 ply)

Figure 3.24 *Protect the pea seedlings from bird predation with wire netting*

Figure 3.25 *Broad bean seedlings ready for transplanting and the tray stood on edge to loosen them*

Figure 3.26 *Planting broad bean seedlings*

Figure 3.27 *Rows of broad bean seedlings*

Figure 3.28 *Watering transplanted broad bean seedlings*

between the canes such that it supports the growing beans in a network. If long sticks from beech trees with plenty of side shoots are available, they will substitute well for bamboo canes. Broad beans produce substantial and heavy leaf and shoot growth, so that it is essential that further lengths of garden twine are put in place as the plants grow. Eventually, the beans will flower and attract honey and wild bees as pollinators, and these can be seen as they move from flower to flower.

Peas are best supported by placing wire or plastic netting around the rows supported by bamboos canes which can be woven through the holes in the net and pushed into the ground, making a palisade or network which will support the peas as they climb up the inside. Initially, the tops of the wire or plastic netting can be tied together, safeguarding the plants from pigeon damage.

Both peas and broad beans will require water if there are prolonged periods of drought.

After transplanting and other husbandry operations are completed, walking boards should be removed and as a matter of routine the ground on which they sat lightly forked over removing ground compaction. This ensures that rainfall can soak into the soil and drain easily and that there is adequate gas exchange between the soil and the atmosphere.

e. Direct sowing

Direct sowing permits sequential harvesting because some varieties require longer growing seasons compared with others. Additionally, some broad bean and pea varieties are winter tolerant and may be sown in mid- to late autumn. This provides very early crops. Sowing dates will differ between southern and more northern areas. Local advice is valuable guidance for sowing dates in particular localities. Gardeners should avoid sowing too early since that results in excessive growth which can be damaged later by excessive frosts or by snow lading. These crops should be protected by cloches or wire netting from pigeon predation since they are particularly attractive especially in hard winters. Broad bean varieties such as 'Claudia Aquadulce' or 'Bunyards Exhibition' and peas such as 'Meteor' or 'Douce Provence' are suitable for autumn sowing.

Spring sowing also provides opportunities for using sequences of varieties. Early peas include the varieties: 'Kelvedon Wonder', 'Jaguar' or 'Premium'; main cropping is achieved with varieties such as 'Alderman', 'Hurst Green Shaft' and 'Little Marvel'. Early season crops of broad beans may be achieved with varieties such as: 'The Sutton', 'Express' or 'De Monica' and for later crops with 'Masterpiece Green Longpod'.

Figure 3.29 *Sowing broad bean seed directly in the garden*

Seed of beans and peas can be sown directly into the garden if no greenhouse or other protection is available. Direct sown crops germinate and mature more slowly than transplanted crops. A sequence of crops can be achieved by planting an early crop using transplants followed by later ones obtained by direct drilling seed into the garden (see Figure 3.29).

Similarly, early, mid-season and maincrop varieties can be selected, providing continuing availability for the kitchen. Surpluses can be deep-frozen for future enjoyment.

Prepare the soil as before, working from a walking board. Rake the soil vigorously, producing a fine tilth and incorporate some "GrowMore" fertiliser at 30 g per square metre. Using the garden line establish where the rows of peas and broad beans will be sown. Then with the rake take out a trench which is the width of the rake head and 5 to 8 cm deep, heaping the soil that is removed alongside the trench. Remove the pea or broad bean seeds from their packet and place in a plastic saucer for ease of handling. Pick up 5 or 6 seeds and sprinkle them into the bottom of the trench, ensuring that each seed is well spaced from its neighbours by about 3 cm each way. Work carefully down the trench, filling it with either pea or broad bean seeds evenly spaced along the entire length. Once the seeds are in place, water the trench well using a can fitted with a fine rose. Then move the soil from the side of the trench over the seed, covering them evenly and without disturbance. Sprinkle slug bait over the trench, which should be marked with 4 short bamboo canes, 1 placed at each corner of the trench. Cover the trench with a low cloche shaped from wire or plastic netting to deter pigeons and other vermin (see Figure 3.30).

Figure 3.30 *Covering rows of directly sown broad bean seed with glass barn cloches*

Monitor the rows for signs of mice being attracted to the seed. If these appear, then trapping will be required (see Figures 3.31, 3.32, 3.33, 3.34 and 3.35).

Figure 3.31 *Comparison between broad bean plants that have been protected by cloches and those that have not*

Once pea and broad bean seedlings have emerged, they will require support and protection as described for transplanted ones. Canes should be placed around the beans with a network of string providing support around the edges of the crop and through the midst of the row. Peas will need wire or plastic netting around them. Subsequent care is similar to that for seedlings raised in the greenhouse and transplanted.

Follow the progress of the peas and beans as they grow, flower and produce pods (see Figure 3.36).

Figure 3.32 *Support network of string and bamboos canes needed for developing broad beans*

Figure 3.33 *Root of young developing broad bean plant; note the long tap root and upper spreading roots*

Figure 3.34 *Developing pea seedlings grown from directly sown seed*

Figure 3.35 *Vigorous broad bean plants protected from bird damage by nets*

Figure 3.36 *Broad bean plants in flower*

Figure 3.37 *Developing pods on broad bean plants*

Make sure that they are receiving adequate water either as rain or from irrigation. Both crops require copious soil moisture for pod forming and filling (see Figure 3.37).

Broad beans also attract some less helpful insects such as black aphids (*Aphis fabae*). These are attracted to the lighter green of the tips of shoots of broad bean. Consequently, removing the shoot tips and 2 to 3 large leaves when bean pods have formed along the length of the broad bean stem after about 10 days helps avoid attacks (see Figure 3.38).

Figure 3.38 *Tip of a broad bean shoot cut off to prevent infestation by black bean aphid* (A. fabae)

Neither of these crops is seriously vulnerable to attack by other pests and diseases provided they are harvested before maturity. As broad bean plants mature, the foliage may become speckled with brown spots; these are caused by chocolate spot (see Figure 3.39).

This disease looks worse than it is and should not cause much concern to gardeners provided the crops are harvested while still young. Mature broad bean plants may be defoliated by this fungus.

Pick the pods as soon as the beans and peas are large enough for eating (see Figure 3.40).

Figure 3.39 *Chocolate spot disease* (Botrytis fabae) *on broad bean leaves*

Figure 3.40 *Harvested and podded peas*

If left too long, they will become over-mature and lose flavour. Over-mature peas and broad beans become dry and "starchy" and are not pleasant eating.

f. Runner beans and dwarf French beans

These are two further legume crops which are very useful and enjoyable. Both are susceptible to low-temperature damage and consequently, require protection if sown directly in the open before the risks of frosts have passed. Both crops are best sown in large trays of compost as described for broad beans and peas (see Figures 3.41, 3.42, 3.43 and 3.44).

Figure 3.41 *Runner bean seed*

Figure 3.42 *Runner bean seed sown in a large tray of compost*

Figure 3.43 *Climbing French bean seed*

Figure 3.44 *Climbing French bean seed sown in a large tray of compost*

Figure 3.45 *Climbing French bean plants ready for transplanting into the garden*

Bean seedlings are very attractive to slugs even in a greenhouse so that it is useful insurance to scatter some deterrent pellets on the surface of the compost as germination begins.

These plants should be planted in the garden as soon as they reach 5 to 6 cm in height and have developed primary leaves with shoots expanding above them (see Figure 3.45).

Dwarf beans are planted into a row of soil which has been well raked and any developing weed seedlings removed. The transplanting process is similar to that described for peas. Dwarf or green beans are a very worthwhile crop and should produce pods within 6 to 8 weeks of transplanting. Varieties available are of two types, either those which form low bush-shaped plants which grow without support or climbing green beans. These require support which is best provided by placing a garden net fixed onto bamboos canes 1.5 to 2.0 m in height (see Figure 3.46).

Figure 3.46 *Climbing French bean plants supported on nylon netting*

Bean pods will form from the base of the plants and then continue being produced up the green wall which develops on the net. Bees find their flowers attractive. In dry weather, keep these plants well watered. Pick the pods as soon as they are a size that is easily (see Figures 3.47 and 3.48) handled, and do this frequently as it encourages continued cropping.

It should be possible to pick over a row of dwarf beans for 6 to 8 weeks. If these are

Figure 3.47 *Climbing French bean plants with pods*

Figure 3.48 *Harvested pods of climbing French beans*

popular in your kitchen, then it will be worthwhile sowing several rows spaced in time achieving sequential maturity and harvesting. Green beans can be deep-frozen for autumn and winter use if the volume of cropping exceeds current demands from the cooks. They are one of the most easily preserved crops.

Runner bean seedlings should be handled in a similar manner. It is essential that they are transplanted before the bean stems begin extending (see Figures 3.49, 3.50 and 3.51).

Runner beans will require a trellis for support which is at least 2 m tall, this is made from long bamboo canes (2.0 to 2.5m long). Cultivate the ground where the beans are going to be planted with the rake and add "Grow-More" fertiliser. Place canes in a row along

Figure 3.49 *Runner bean seed germinating in compost*

Figure 3.50 *Runner bean plants ready for transplanting into the garden*

Figure 3.51 *Runner bean plant and its root system*

the garden line spaced at 20 cm apart, these should be pushed into the soil to a depth of about 30 to 40 cm. Runner beans require stout, wind-resistant support because they produce a very heavy canopy of stems, leaves and beans as the season progresses. Transplant two bean plants at each cane station. Now insert a second row of bamboo canes each opposite an existing one. Place a cane along the top of the row of canes so that it sits in a cleft formed by crossing the tops of the two opposing canes. Firmly tie in the horizontal cane at about 4 to 5 cm from the top of the vertical canes with 5-ply garden twine. Place two runner bean plants at the base of each of the second row of vertical canes. All the plants should be well watered and slug deterrent scattered round them.

The plants will produce long, extending stems growing in a clockwise spiral, which seek out the canes and climb them by growing around them quite naturally (see Figure 3.52).

Occasionally, a misguided stem may require help by being tied into a cane. These plants have few enemies apart from the black bean aphid (*A. fabea*), which should be eradicated with a suitable insecticide. Caution is needed when using these materials as runner bean flowers are very attractive to bees. The best policy is to apply insecticide very early in the mornings or late at night on cool overcast days when bees are least likely to be active.

Runner beans crop continuously from mid-July to the end of October or until the first frost. Pick the pods as soon as they are of a suitable size for the kitchen, surpluses can be

Figure 3.52 *Runner bean plants climbing bamboo canes and flowering*

deep frozen for future use. As with other legumes, runner beans are best harvested when they are young and juicy well before the seeds have started swelling. Maturing runner bean pods contain strengthening tissues that reduce their tenderness and eating quality.

Underpinning knowledge

3.1 Reproduction asexually and sexually

Reproduction has two functions, firstly, as the basis for processes by which plants grow and expand and secondly, making provision for producing future generations.

The plant kingdom contains an array of organisms of differing complexity living either wholly or partly in water or on the land. In the seas and waterways, there are forms of single-celled algae and multicellular water-weeds and sea-weeds. Among the earliest plants that colonised land were ancestral forms of present-day liverworts, mosses and ferns, which to varying degrees still require very wet or at least moist habitats. Later in geological time, the early conifers, Gymnosperms, evolved lifestyles capable of living in drier conditions. Eventually, the flowering plants, named Angiosperms, which now predominate in our gardens, developed with more complex lifestyles. Great diversity developed in these lifestyles, some, such as the cacti, capable of living in very dry, desert conditions (see Figure 3.53) while others, such as water lilies, resumed aquatic existences (see Figure 3.54).

Each level of evolutionary complexity in plants from the liverworts to the Angiosperms has slightly differing modes of sexual reproduction.

Naming plants

Gardens mainly contain flowering plants, collectively known as Angiosperms consequently, only their details are dealt with here. There is huge variation in the structure of flowers produced

Figure 3.53 *A display of cacti in bloom*

Figure 3.54 *A display of water lilies in bloom, grown in a container and suitable for a garden patio*

by the Angiosperms. This complexity perplexed the early botanists, mostly monks and herbalists making medical use of plants. In their naming systems for explaining and recording plants, they used extended Latin descriptions which at that time lacked a logical scientific basis. During the early 18th century, two botanists, an Englishman John Rae and more importantly the Swedish researcher and traveller Carl Linnaeus (see Figure 3.55), realised that the structure of flowers provided key information for identifying and rationally naming plants with a logical scientific basis.

Figure 3.55 *Bust of Carl Linnaeus*

Linnaeus' studies of plants sent from around the world provided him with scientific facts, from which he devised a simple binomial system for naming plants based on floral characteristics. It is called a "binomial" system because the first name, "genus", identifies a closely related group of plants having similar but individually distinct features and which is generally unlikely to reproduce compatibly with each other. The second name, "species", describes a smaller sub-group within a genus whose members are capable of sexual reproduction with each other. Above the genus level are hierarchies of larger groupings into families, orders and phylla.

Floral structure has been the essential bedrock for defining plant classification since Linnaeus' inspirational studies. Each combination of genus and species defines a specific plant type and its initial formal description is published in a suitable scientific taxonomic journal in Latin. That retains the conventions of the early herbalists and is universally acceptable and understood. The specimen from which that description is made is placed in an herbarium as the "lectotype", or reference type, which may be consulted by future generations of botanists and horticulturists. Reference collections are held in internationally accredited herbaria, such as those at the Royal Botanic Gardens at Kew and Edinburgh. These herbaria hold collections accumulated over centuries of plant hunting worldwide. Herbaria are of increasing importance because their plant material is a biological library containing detailed botanical information which only now can be re-interpreted by molecular analysis and is leading towards a considerable review of the naming of flowering plants. Herbaria of cultivated plants are held by horticultural societies mostly in Europe and North America. A leading international collection is at the Royal Horticultural Society, Wisley, Surrey.

Asexual and sexual cell division

Cell division and reproduction presents all organisms with two essential but opposing requirements. Firstly, cells must divide in order that organisms may grow. Instructions for growth, fruiting and maturity are stored in the nuclei of cells. Nuclei are the largest and visually most obvious organelles in cells housing a library of genetic instructions, which controls all aspects of plant development from seed germination through to senescence and death. Nuclei were first observed by the English botanist Robert Brown, nucleus is Latin for kernel or nut. Genetic instructions are contained in a series of chain-like structures known chromosomes, they were first identified by the German scientist Walther Fleming in 1882. The number of chromosomes varies between different species and sometimes within species.

When parent cells divide into two progeny (daughter) cells, these must be equipped with identical sets of genetic instructions for growth, maturation and the formation of the structures by which reproduction is achieved. Unless successive generations of daughter cells have these identical sets of genetic instructions uniform, coherent growth is impossible. This process is referred to as asexual reproduction. The process of cell division that achieves this is called mitosis (Greek *mitos* meaning "thread") (see Figure 3.56).

Gardeners make use of this characteristic by developing groups of plants equipped with the same genetic instructions; these are known as clones and are obtained by methods of asexual propagation, as described in Chapters 1 and 8. They result, for example, in identical potato tubers of specific varieties, such as those listed in Chapter 2. The implications of mitosis in the garden are explored further in Chapters 5 (fruit), 6 (bulbous plants) and 7 (flowering plants).

Plants must also reproduce sexually; this is achieved through pollination, flowering and seed production. Sexual reproduction introduces genetic variation in succeeding generations. Variation between generations is essential because it allows plants, and all other organisms, opportunities for adaptation as environmental conditions change. During sexual reproduction, the chromosomes divide in a manner which introduces variation; this process is known as meiosis (see also Figure 3.56).

Mitosis

Mitosis is a simple one-stage division process which replicates existing structures. Each nucleus of a green plant contains several chromosomes, the number varies with different organisms. Some plants contain several copies of each chromosome, this is known as polyploidy and contributes to added vigorous growth and flowering. Division and replication processes during mitosis are known as the "cell cycle", during which a series of coordinated changes take place in both the nucleus and in the cytoplasm of the cell which surrounds the nucleus. Each of the phases of the cell cycle progresses seamlessly and continuously.

For most of the time, cells are growing and maturing without the chromosomes in the nuclei being visible, even under a microscope. When mitosis starts, the nucleus begins building replicas of the genetic material (see again Figure 3.56). At this stage, the chromosomes become visible, under a microscope, within the nucleus. Each chromosome divides into two equal halves known as chromatids. The chromatids become aligned along a central equator. This line is not a physically obvious structure but is indicative of the plane in which cell division occurs. Guidance for the separation and movement of chromatids is achieved by a series of microtubules or ropes. These are gathered together on either end of the membrane which surrounds the nucleus. Orientation of these tubules determines the direction in which the whole cell divides and that controls the plane in which the subsequent new cells will grow and hence the structure of the subsequent root, stem, leaf, flower or fruit. The microtubules do not actually contract but

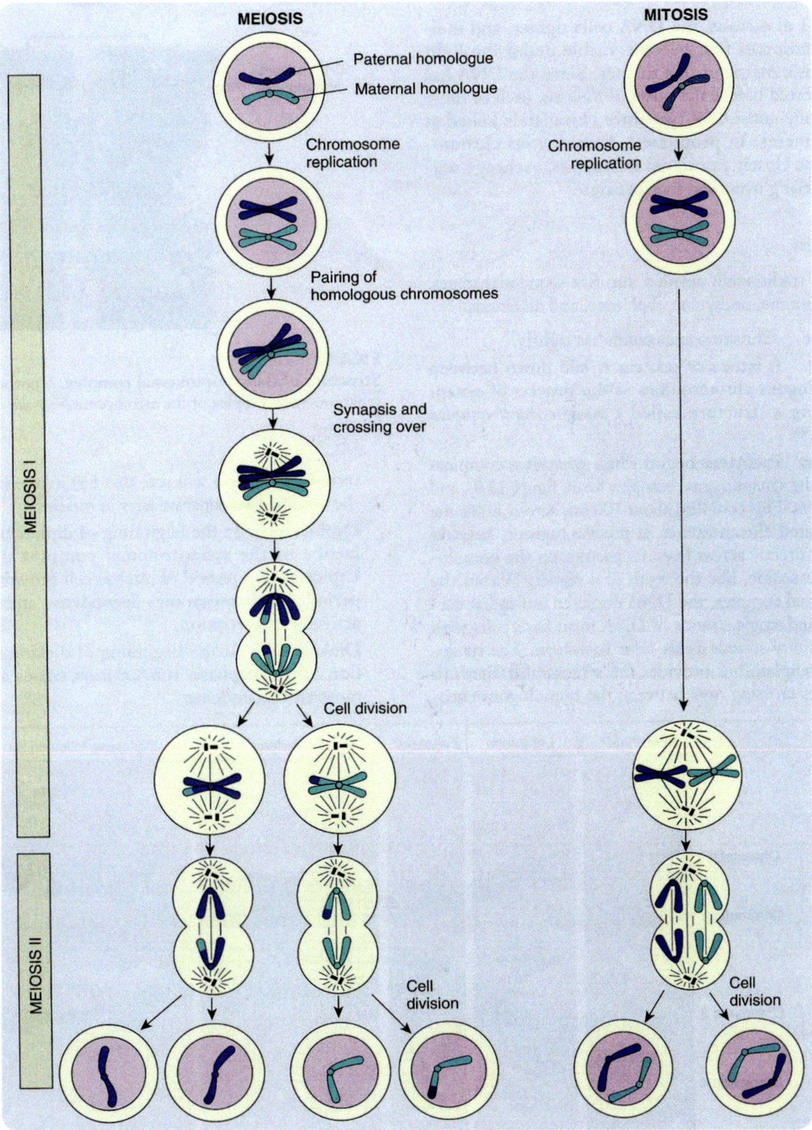

Figure 3.56 *Comparison of the processes of mitosis and meiosis. In* meiosis *note the exchange of genetic material between chromosomes*

shorten by the loss of sections in their structure. Once the two equivalent sets of new chromatids have drawn sufficiently far apart, nuclear membranes reappear. At this point, each set of new chromosomes is sealed separately within their own membrane and can function as the new instruction book. The normal work of genes on the chromosomes recommences once two separate nuclear membranes are established. Each gene is regarded as the smallest physical unit of heredity and is the set of instructions for a specific product or process.

At this point, the cell itself divides into two daughters. Each cell then contains a new nucleus fitted with an identical set of chromosomes equipped with genes derived from the parent cell. The result is two new identical cells which can grow, mature and then themselves divide. They also have the capability of developing into any of the range of specialised cells contained within a plant.

Each chromosome is composed of a substance known as chromatin, this is a complex of 60 per cent protein and 40 per cent deoxyribonucleic acid (DNA). Much research throughout the latter half of the 20th century, and continuing now, has revealed how the instructions contained in DNA control and direct every facet of an organism's life and death. The chromosomes are strings of individual units, genes, each of which is part of an information storage and retrieval system. This system contains many thousands of instructions for germination, growth, reproduction and senescence of an organism. DNA is a double-stranded fibre formed in a double-helical shape, as identified by the Cambridge biologists Francis Crick and James Watson in the 1950s, and informed by Rosalind Franklin's pictures of the crystalline structure of DNA.

Asexual cell division can be as quick as 20 minutes in actively multiplying tissues such as stem and root meristems, or requires several hours in older tissues. In plants, some of the cell membranes contain cellulose forming a rigid wall. This is typical of the plant cells found in stems and parts of roots. These cells are strengthened but still have the capability for flexing and bending in wind currents.

Flowering plants are coordinated multicellular organisms; consequently, division and multiplication are regulated throughout the whole organism and organised by series of growth regulatory substances (see Chapter 8).

Meiosis

Meiosis is a reduction division which introduces variation between its products (see again Figure 3.56). Meiosis provides variation in the characteristics of organisms which as Charles Darwin's and Alfred Wallace's researches in the 19th century recognised results in the evolution of new species. Some new species will have capabilities for coping with shifts in environmental conditions in what has been termed the "survival of the fittest". Plant breeders use these properties for variation when producing new varieties of plants for commerce and in the garden.

The formation of sexually reproductive cells takes place in specialised organs which are the central parts of flowers. These appear at the ends of shoots and result from a change from vegetative to reproductive growth triggered by growth-regulating substances. The end-product of sexual reproduction in flowering plants is the development of a new generation of seeds which vary to lesser or greater degrees from their parents.

Sexual organs

Generalised flower structure consists of four parts growing from the top of a shoot which is known as the receptacle. From this develops a series of leaf-like structures, sepals, collectively forming the calyx (see Figure 3.57).

This is a largely protective structure which prevents damage to the more delicate structures growing within the calyx. In many plants, the calyx is green and functions photosynthetically.

Figure 3.57 *Generalised flower structure*

Within the calyx is a ring, or whorl, of petals which are frequently highly coloured and varying in size and shape. In some plants, the calyx and petals are not clearly differentiated from each other, for example in tulips (Chapter 6), and are known as tepals. The function of petals and tepals is largely the attraction of pollinators. Plant breeders have over centuries selected petals and tepals in an almost limitless range of colours, shapes and sizes, building on what had been produced naturally during evolution in the wild.

Within the petals are the working sexual organs. Firstly, the male parts, or stamens, consist of stalks on the top of which are anthers. Pollen cells, grains, are produced in the anthers and these are the male contribution for reproduction. Usually at the centre of a flower is the female organ, known as the pistil. This consists of an ovary at the base which is essentially a "womb" where the new generation of seeds develop; above this is a stalk-like style carrying a stigma or receptive surface at the top. This is where pollen grains from the male organ are deposited in a process known as pollination.

The sexual process

Meiosis takes place inside each of the male and female sexual organs producing cells known as gametes. Meiosis in the gametes is a two-stage process in which the parts follow each other sequentially and seamlessly. When the gametes unite following fertilisation in the ovule, genetic material from both parents is brought together, in what is known as the diploid state. Meiosis functions as a reduction mechanism (refer again to Figure 3.56). This reduces the diploid state in half and consequently the gametes are in, what is termed, the haploid state. Without this reduction division from diploid to haploid cell nuclei would be very rapidly over-loaded with genetic material.

At the start of the first stage of meiosis, the chromosomes become visible within the nucleus as happens in mitosis. In all cells, there are pairs of chromosomes derived respectively from the male and female parents, these are referred to as paternal and maternal homologues.

Each homologue consists of a series of instructions (genes) received from the respective maternal or paternal parent. Each of these homologues divides forming four structures known as chromatids. The pairs are physically closely associated in a process known as synapsis. At this point, variation is introduced by crossing-over (exchange) of genetic material between each of the chromatids originating from the male and female parents. During crossing-over there is active exchange of genetic information between the chromatids derived from the maternal and paternal chromosomes (see again Figure 3.56). Consequently, genetic information previously present in one chromatid is moved to another. This may result in only tiny changes but these are sufficient to provide divergence from the original parental material. There follows a total dissolution of the nuclear envelope, and the pairs of chromatids move towards the opposite sides of the equatorial plate. They are drawn apart towards the poles of the original cell by the action of microtubules.

There may then be either a period quiescence in the two cells or they move directly into stage two. This stage is similar to mitosis (see again Figure 3.56). Each of the four chromatids have been separated in stage one into two nuclei. Each member of the pair now separates moving towards different ends of the nucleus. Finally, a nuclear envelope forms separately around each of the four new sets of chromatids some carrying altered genetic instructions. These now function as four new sets of chromosomes in the haploid gametes. Some gametes contain a slightly different genetic constitution from the others and from the original parents. The diploid state will only be restored following the union of two gametes, male and female, during sexual fertilisation. Genetic mixing, known as recombination, achieved by crossing-over is the foundation for future variation from the parental genetic composition.

The variation which is established by meiosis may be advantageous for the subsequent individuals and generations or it may be a disadvantageous. The former will be exploited by the individual as it provides enhanced fitness as capabilities for withstanding environmental change. Disadvantageous genetic material becomes an evolutionary dead-end and the individuals die out. This is the processes of "survival of the fittest" as described by Charles Darwin and his colleague Alfred Wallace.

Pollination and sexual fertilisation

In flowering plants, the pollen grains carry the male contribution of genes, and in the process of pollination these are transported to the female stigma. On the stigma, the pollen grains germinate, forming a pollen-tube down which the male reproductive cells travel into the ovary within which are female cells, or ovules. Union of the male and female cells reconstitutes the diploid gene complement of the new plants. In this process, known as fertilisation, there is further mixing of genes from the male and female parents as the haploid gametes unite into a diploid cell and nucleus. Consequently, seed formed from fertilised ovules contains a slightly different genetic complement to their respective male and female parents. The seeds that result from fertilisation develop in the ovary from a series of ovules. Surrounding each ovule is a seed coat, or testa, and accompanying the developing embryo is a structure, the endosperm, which supplies it with nutrients. Individual seeds may be contained within a dry pod or a juicy fruit where one or more individuals develops.

Pollen and seed dispersal

Pollen movement and seed dispersal in flowering plants have evolved using either intricate relationships with both physical environmental processes or biological partnerships with animals (see Figures 3.58, 3.59 and 3.60).

The generalised model of flower structure (see again Figure 3.57) is hugely modified across the enormous range of Angiosperm plants.

Pollination may utilise the physical processes of wind and water or require the carriage of pollen grains by insects, birds and other animals. Insects, particularly bees, are very important agencies for pollen movement. They move industriously across flowers seeking out those where pollen is mature and ready for transfer. Frequently, they are attracted to flowers by their nectaries; these are small nodes of cells forming a gland usually growing at the base of the corolla. These exude liquids rich in sugars, which bees and other insects find very attractive. Some

Figure 3.58 *Bee pollination of grape hyacinth* (Muscari armenaicum) *flowers*

Figure 3.59 *Wind pollination of grasses (*Setaria *spp.) in flower*

Figure 3.60 *Hazel (*Corylus *spp.) with typical wind-pollinated yellow male catkins which produce pollen*

flowers have very highly modified structures moulded by evolution which leads to great specialisation in the animals which are able to collect and transport pollen.

Inanimate or animal agencies are also enlisted by flowering plants as vehicles for moving seed, seed capsules and juicy fruits. Seeds surrounded by juicy fruit tissues, as with strawberries and raspberries, or with succulent vegetables such as marrows are eaten by birds and other animals, such as rodents and dispersed in their droppings (see Figure 3.61).

Figure 3.61 *Seeds formed in the succulent fruit shown in lateral and transverse sections of a marrow*

Alternatively, seed may adhere onto the coats of animals as may be the case with the ancestors of modern beetroot. Several individual seeds are combined into a dry fruit which has spines capable of attaching on to a convenient carrier (see Figure 3.62).

Since wild beet (*Beta maritima*) has habitats on sea coasts, it is also possible that these complex dry fruits are sufficiently robust that they could be transported by water from one place to another.

Plants where wind is the agency for pollen dispersal or seed movement usually also have highly modified flower structures. In pollination, the anthers extend well beyond the corolla and this allows wind currents to collect and disperse the pollen onto naked and extended stigmas. Similarly, wind transport of seed requires the formation of huge quantities of very light weight seeds released possibly explosively from the ovary or capsule. In some cases, simply the drying out of seed cases releases their contents explosively, as with the pods of brassicas or poppies as a physical dispersal mechanism. The fine seed may then be carried on air currents (see Figure 3.63).

Figure 3.62 *Beetroot seed head: a complex of individual seeds*

Figure 3.63 *Explosive seed dispersal from a poppy (*Papaver *spp.) capsule*

Dispersal is an essential mechanism, ensuring that populations are not concentrated in one area. There is a continuous need for plants moving from existing into new habitats. Although individual plants are "rooted" in one place that does not mean that their populations are immobile and incapable of colonising new niches. Each seed dispersal mechanism which has evolved promotes seed dissemination and eventually germination at a distance from the original parent plants. The reason for pollen and seed dispersal is that they ensure that plants increase the areas which they colonise, entering new habitats with fresh nutrient supplies.

The underlying functions of flowering are that it permits genetic mixing within plant populations, thus increasing diversity and the extension into new areas for colonisation. The varying mechanisms evolved for pollination and fertilisation increase opportunities for expanded diversity and hence capacities for coping with changing environments. In the processes of evolution, plants have diversified their floral structures such that they may be either self-pollinated, for example potatoes, or cross-pollinated, for example runner beans. The processes described so far largely relate to cross-pollinating plants where there is a very high degree of genetic mixing. In self-pollinating types, pollen is transferred directly within the flower from anthers to stigma possibly without the need for opening of the corolla of petals. Self-pollination can be highly advantageous for plants which grow in very specialised habitats such as high mountainous elevations, sub-Arctic conditions or where there are few other plants in colonising situations. The resultant plants offer considerable challenges for gardeners in cultivation because of their need for specialised care. Specialisation may be a big advantage for plants living in a particular niche environment but if conditions change it becomes threatening.

An alternative evolutionary pathway is found in plants which are fertilised by out-crossing. This produces highly diverse plant populations but does demand the presence of either physical or biological mechanisms for the transfer of male cells to the female. Plants which out-cross may produce either male or female organs on different plants (known as dioecious, "two houses) and that ensures that added genetic mixing takes place before successful seed production happens. The gardener will find some of these plants challenge his or her skills because in cultivation there is a need for both male and female plants sited in close proximity if seeds and fruit are going to be produced successfully in the garden. A classic example of this is holly (*Ilex* spp.) where both male and female plants are usually required in order that berries are available for Christmas decorations. Some out-crossing plants have both male and female organs in the same flower (known as monoecious, "one house"). The transfer of pollen outside these flowers happens because either the male or female organ ripens before the other. Consequently, one of the pollination agencies is still required, ensuring pollen transfer and viable seed production.

3.2 Across the generations

Seed is the means by which flowering plants propagate from one generation to the next. Seed germination and the growth of seedlings is the first stage in the establishment of robust and high-quality plants.

As described in Underpinning knowledge 3.1, seeds are formed following the processes of pollination and fertilisation leading to the production of an ovule encased in an ovary, which is an embryonic plant. Alongside the ovule is an endosperm which provides the developing seed with nutrients in the earliest stages of growth. Both are encased in a membrane or seed coat, known as the testa. Once they are mature, seeds may enter a period of dormancy from which the gardener stimulates growth. Seed bought from garden centres or mail-order companies is held in a dormant state in the packet.

Seed production

Most seeds are harvested from specialist crops grown for that purpose. Once seed crops are harvested, they are dried and passed through sieving and shaking machinery which removes remnants of plant material and soil particles generally cleaning the seed. It is then packed into aluminium foil envelopes during which process the humidity surrounding the seed is reduced. Seed can then be stored in good condition for several years at low temperatures. Since crops can be very productive, vegetable and flower seed is grown in large quantities and is carried over for several years. There are strict regulations, however, which control the quality of seed, especially that of vegetables. Each type of seed must meet predetermined conditions of purity and germination before it can be sold. Samples of seed of existing and new varieties are tested by trained seed analysts under laboratory conditions, establishing that they meet minimum standards of germination, cleanliness and trueness to type. Additionally, in Europe the variety sold must reach specified standards. Registered vegetable varieties must pass officially regulated tests for Distinctness, Uniformity and Stability (DUS), meaning that they are distinct from other comparable types, the crop produced will be uniform and without gross variations and the characteristics are stable over several cycles of propagation. Adherence to these criteria especially for vegetable seed is signified on the backs of seed packets with statements as to purity and the date by when the seed should be used. Flower seed growers normally also carry out substantial testing programmes, ensuring the cleanliness and trueness to type of their products. The United Kingdom authority responsible for administering variety regulations is the National Institute of Agricultural Botany (NIAB), Cambridge, contracted by the Government's Department for the Environment, Food and Rural Affairs (Defra).

For some very highly prized ornamental garden plants, there is an independent system of registration and monitoring arranged by collaboration between authoritative internationally recognised societies. Independently, some organisations such as the Royal Horticultural Society (RHS) provide their members with seed taken from interesting and unusual ornamental plants. This is one way in which gardeners may increase the diversity of their plantings.

Some varieties of fruit, vegetables and flowers carry indications that they are registered for plant breeders' rights (PBR) in Europe or have plant patent rights if they are of North American origin. These statutory systems allow plant breeding companies rights for the recovery of fees (royalties) for multiplication of these varieties. This protects the intellectual property of plant breeders and their companies who have invested many years of work and finance in producing improved varieties. Rose breeders, in particular, have developed a very tightly controlled system for registering, recording and protecting new varieties, noting particularly pest and pathogen resistance, floriferousness, growth pattern and fragrance.

Plant-breeding companies use multiplication centres worldwide where the environmental conditions are optimal for the growth and flowering of particular plants with an absence of pests and diseases which might reduce seed quality. Frequently, this means that crops are grown in warm temperate or tropical countries. Some very specialised seed, such as hybrids of tomatoes and cucumbers, are grown in glasshouses in more temperate regions. For these crops, pollination may be done by hand, ensuring effective and reliable fertilisation; they are also carefully harvested manually. Consequently, the seed is expensive and each packet contains only a few individuals.

Figure 3.64 *An example of plant breeding in the variation in root shapes achieved in varieties of 'Daikon' radish*

Breeding new varieties is a long-term and expensive process, frequently taking more than ten years. Incorporating resistance to pests and pathogens, which is a highly prized characteristic, for example, requires assembling detailed understandings of the plant and its adversaries. Where this is achieved satisfactorily, the seed company will have an exciting new "product" for marketing (see Figure 3.64).

Seed packets are a means for colourful marketing, displaying their contents to the best advantage. The suppliers of seed may advertise and market their products using several brand names which they anticipate will increase gardeners' interest.

Seed quality

Most vegetable varieties sold to gardeners and many flowers are now labelled as F_1 hybrids. This indicates that the plants produced from the seed will be particularly uniform and deliver reliable yields. Some gardeners may wish to grow older heritage types of varieties because they believe that these produce superior flavour or have some other interesting characteristic. These are out-breeding varieties, which are much more variable in yields and quality compared with F_1 hybrids. The costs involved in producing F_1 hybrids are higher than for open-pollinated varieties, and hence the seed is more expensive.

Seed germination

Chapter 1 describes the conditions which are necessary for seed germination. These are supplies of water, warmth, oxygen, a substrate and nutrients. Once seed is sown, water from the environment penetrates through the seed coat, or testa. Under favourable temperature conditions, oxygen is also taken into the seed and, as a result, supplies of nutrients stored in the seed as complex carbohydrates and fats are broken down, following energy release during respiration, under the influence of enzymes. This produces further sources of energy which the emerging seedling uses for cell division and growth. Germination is first indicated by the emergence of a tiny root protruding through the seed coat. This root, otherwise known as a radicle, extends into the substrate and starts forming growths from its surface cells known as root hairs (see

Figure 3.33). These form an absorptive network through which the developing seed imbibes water from the substrate and supplies of mineral macro- and micro-nutrients, as described in Chapter 2. Research shows that at this early embryonic stage the root and its hairs begin interacting with microbes in the soil. Many of these microbes are beneficial for the growth of the root, providing defence mechanisms against other more damaging microbes.

Seedling growth

As the root, radicle, develops, it starts delivering water and nutrients to a shoot, plumule, which is forming at the other end of the embryo. This emerges from the seed and grows upwards towards the soil surface. Various biological stimuli ensure that the plumule grows towards light (see Underpinning knowledge 7.1). Once emerged above soil level, the plumule manufactures chlorophyll and photosynthesis commences. The seedling has then become an independent entity capable of manufacturing complex compounds from atmospheric carbon dioxide. At that point, the root and shoot commence a partnership which results ultimately in a mature floriferous and fruitful plant. Stimuli which encourage germination relate to the environments in which plants have evolved. Some seeds will germinate almost immediately after they have formed, for example groundsel (*Senecio vulgaris*) (see Figure 4.63). These are plants which are rapid colonists capable of taking over ground very quickly. They include some rapidly growing weeds. Often these produce several generations of seedlings in one season and are known as ephemerals. By comparison, woody shrubs and trees often produce seeds which will not germinate until they are exposed to a period of winter chilling. This is a protection mechanism that ensures that the seed does not germinate and produce young vulnerable plants as winter conditions set in.

Seedling vigour

Vigour describes the rate of growth of seedlings, their robustness, health and capacity for producing acceptable crop yields. Gardeners should provide favourable conditions for germination and growth of young seedlings (see Figure 3.65).

But germination may vary considerably between batches of seed even where these are of the same variety. Lettuce seed for example is harvested from a "flower" which consists of several individual flowers or florets. The seed from each of these flowers matures at slightly differing rates and as a consequence will vary in the speed of germination and early growth.

Figure 3.65 *Uniformity of germination achieved with lettuce seedlings*

Vigour is measured by the rate at which seed during the early stages of germination imbibes water from its environment and the effectiveness of enzymic processes that drive the release of energy from stored reserves. These processes result in the initiation and development of the radicle and plumule and their emergence. Uniform and vigorous seedling growth is a major requirement for commercial growers because it results in efficient development, land use and high-quality produce. Tests which establish the respiratory efficiency of germinating seed, photosynthetic capabilities of developing seedlings and rapidity of transitions between growth stages help establish vigour characteristics for batches of seed.

Stratification

Seed of some plant species enters a period of dormancy which provides natural protection from germinating during winter conditions. Dormancy is overcome (broken) by accumulating a required number of days below a pre-determined cold temperature, termed "degree-days". Requirements for these periods of low temperatures as preconditions for germination may be mimicked by stratifying the seed. Stratification is a horticultural term for the process by which the chilling requirement of seeds, vernalisation, is satisfied. Seed is placed in a polyethylene bag and mixed with some slightly moist horticultural-grade sharp sand, obtainable from garden centres. Place the bag in a refrigerator at about +4 °C. Do not use a deep-freeze, as very low temperatures will kill the germinating seed. Once the seed has been chilled for about 12 weeks, it should germinate when placed in the normal conditions of water, warmth, light, oxygen and a substrate. Alternatively, the seed can be sown into a pot of seedling compost and this placed in a cold frame for the winter. Regrettably, British winters are becoming milder and may not be sufficiently cold in order to stimulate germination. Seed placed in a cold frame should be monitored regularly, ensuring that the pots do not dry out and that there is no predation by vermin such as mice. These animals find seed and seedlings are a useful supplement for their winter diets.

As the shoots emerge, primary seedling leaves, known as cotyledons, may appear. These have formed within the maturing seed soon after fertilisation and form a large proportion of the seed structure placed alongside the endosperm. In some flowering plants, these cotyledons emerge above soil level as the first pair of leaves. In other types of plant, they may only appear as a single first leaf. This identifies one of the most basic characteristics of the flowering plants.

Plants may be divided between those which produce two cotyledons and are known as dicotyledonous (di-cots for short) and those which have single cotyledons and are known as monocotyledonous (mono-cots). The di-cots are broad-leaved plants where the veins in the leaves are seen as a network mesh; the mono-cots have parallel veins in the leaves and are typified by the grass family (*Gramineae*) (see Figures 5.51, 5.52, 5.53 and 5.54). This is an important group of plants since it contains all of the cereals, such as barley, maize, oats, rice and wheat, and hence is a major source of food worldwide. Mono-cots are leafy plants and may be devoid of a main stem and generally forming low-growing plants as in the grasses. The di-cots will produce a central stem with side branches, growing in some cases to 50 metres or more in the case of forest giant trees. This topic is discussed in more detail in Chapter 5 (see Figures 5.51 and 5.52, dicotyledon and monocotyledon internal and external structure).

Growth in seedlings takes place at the tips of roots and shoots where there are layers of cells, known as meristems, which multiply very rapidly forming a generalised layer. From this layer, cells differentiate into the whole range of tissues present in roots, leaves, shoots, flowers and fruits. Differentiation is the property by which an initial unspecialised cell can develop into particular cells with specialised functions.

3.3 Life cycle development

The cycle of life in plants describes how they develop from seeds by germination followed by the rapid growth of roots, stems and leaves; the formations of flowers and fruit completes the cycle by producing seed, at which point senescence and death may follow. Alternatively, the plant may continue these cycles for several years. The cycle of life involving germination, growth, flowering, pollination, fertilisation, seed and fruit development, seed dispersal, aging and death and decay are summarised in Figure 3.66.

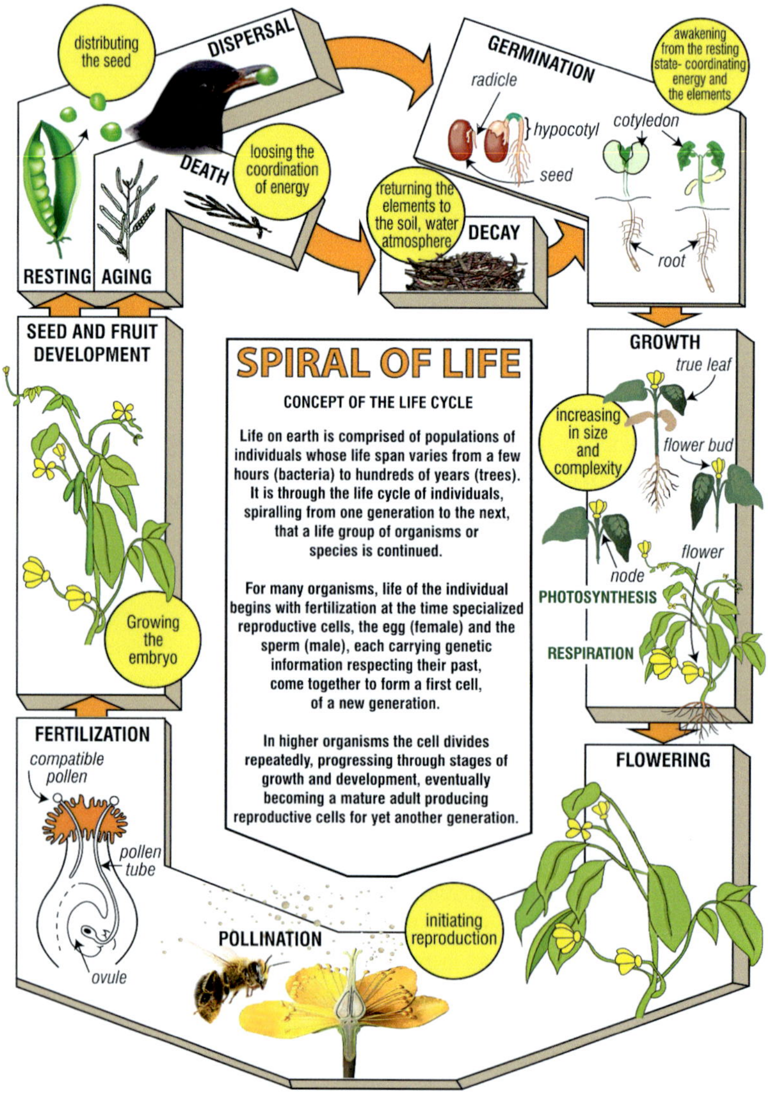

Figure 3.66 *Life cycle of plants*

Inheritance and the environment

The life cycle of plants is described by two terms. Morphology, which describes the physical and visible characteristics of an organism and physiology, which describes why and how growth, flowering and fruiting take place. Two over-riding factors control the morphology and physiology of all organisms. These are the genetic inheritance passed from generation to generation by the process of sexual reproduction (see Underpinning knowledge 3.1) and the environment in which growth takes place. Form and function result from an interaction between genes and their environment. Charles Darwin and Alfred Wallace identified the paramount importance of the environment in shaping the character of each organism through the power of natural selection. At about the same time, recognition of the part played by inheritance in the expression of the outward characteristics of organisms was first explained scientifically by Gregor Mendel, a monk living in the Czech town of Brno in the mid-19th century. Mendel achieved this by a series of careful experiments studying the inheritance of characteristics of peas such as smooth and wrinkly seed coats. From his data, he calculated the number of factors, now called genes, required for the expression of a particular character. He showed that individual characters are determined by genes and passed from a parental generation to its progeny. From this knowledge, the science of genetics developed from the early 1900s onwards, and that has provided a mathematically accurate basis for scientific plant breeding. Originally Mendel published in a local Hungarian journal and this work was re-discovered after 50 years by three botanists, Hugo DeVries, Carl Correns and Erich Tschermak, who verified and publicised it. The result of the re-discovery of Mendel's studies is the huge numbers of excellent varieties of vegetables, fruit and ornamentals which now benefit gardeners.

From the 1950s, Francis Crick and James Watson and subsequent scientists have unravelled the structure of genes and the mechanisms whereby inherited instructions are translated into action within the cells of all plants and expressed outwardly in shape and colour of garden plants. This has culminated in the unravelling of the entire genetic code for important plants. This was first achieved for *Arabidopsis thaliana*, in 2000 by an international consortium of scientists. *Arabidopsis* is a very important plant because it has a very small complement of chromosomes and can be grown easily and rapidly under controlled laboratory conditions and hence is used as a "model" from which the characteristics of more complex food crop plants may be studied and understood. This seemingly insignificant rock garden plant and native weed has become the major model plant for molecular biologists. *Arabidopsis* germinates, grows and reproduces in 15 days; that speed of propagation resembles for botanists what is achieved by the zoologists' model animal, a fruit fly *Drosophila*. Studies of this plant have led onto investigations for major food crops such as wheat, rice, maize, tomato and apple.

Instructions contained in the DNA as codes are sent out from the nucleus out into the cells using a messenger, ribose nucleic acid (RNA) which takes the genetic code to cell organelles known as ribosomes. There the messages are transcribed and then translated forming proteins. These initiate and control very precisely the chemical reactions taking place in cells. The results of these reactions are seen outwardly as the visible characteristics of plants. Equally important are unseen processes resulting from these reactions happening within plants and which are responsible for the processes of growth and reproduction. Overall, the forms and functions in plants result from the movement of DNA codes held in the nuclei and passed on into organelles via RNA in the cells. Each cell can be considered as a smoothly operating biological factory where the control centre, nucleus and its DNA, sends commands to process engineering machinery which either constructs or dismantles products. Recently, the impact of environment on expression of DNA codes has become appreciated scientifically. Acceptance that gene expression can change as a result of environmental influences is a fairly recent scientific development. Previously, such suggestions were discounted sometimes vigorously as heresy.

Growth form and function

Gene expression, the translation into the processes of growth and development and the effects of the surrounding environment commences as the seed emerges from its dormant state and continues in its life cycle. Water is taken into the seed and rehydrates its content. This combined with warmth and oxygen provides the basis for cell development and division leading to the growth and emergence of the root and shoot, as shown in Chapter 1.

Respiration takes in oxygen and degrades the complex carbon reserves held as carbohydrates or oils in the seed, changing them into sources of energy for growth and cell division (see Underpinning knowledge 1.2). Water and nutrients are taken into the root and passed onwards into the developing shoot which emerges from the seed. This physically forces its emergence through the soil surface. Developing shoots respond positively to the weak amounts of light which penetrate the surface layers of soil and possibly negatively to gravity. At that point, the outer cells begin manufacturing chlorophyll and photosynthesis commences. The seedling has now changed from surviving on carbohydrates, fats and proteins provided from the parent plant into a free-living organism. Photosynthesis provides the plant with a supply of complex carbon-based molecules. Some will contain mineral elements such as nitrogen, sulphur, magnesium and iron derived from the nutrients absorbed by the roots. Others will become long-chained or cyclic structures which form reserves of carbohydrates and fats. Alternatively, they are constituents of cell walls required in the building and further development of leaves and stems. Some aromatic molecules which are formed have protective functions providing defence against predatory microbes and herbivorous insects and larger animals. Each of these processes in plants result from instructions present in the genes which are translated within the cell into a myriad of cascading chemical reactions which form messages and messengers triggering the life cycle of a plant.

The plant responds both to the presence of light, day length and night length (see Underpinning knowledge 7.1). Plants grow towards a light source, responding by increasing cell division in the apices of stems and shoots (see Figure 3.67).

Figure 3.67 *Comparison of mustard seedlings grown with unilateral light (left) and all-round illumination (right)*

Plants grow by forming new cells following the division of parental ones in mitosis. These expand and elongate forming tissues such as the photosynthetically active cell layers in leaves and stems. Alternatively, they may form the active transport cells in the outer layers of the vascular tissue or they form flowering and fruiting structures. This is where plants become valuable for gardens. Crops grown for stems and leaves result from an expansion of the early formed leaves and from the development of vegetative buds increasing plant size, as for example lettuce, leeks and many of the cabbage family. Root crops such as potatoes begin forming new tubers as the foliage thrives and transports carbohydrates from the leaves, which boosts their formation. This applies also for perennial crops such as asparagus and globe artichokes. Day and particularly night-length, influence the onset of flowering and reproduction in plants originating from temperate regions (see Chapter 7). Plants from tropical regions may grow and mature as a result of other stimuli such as temperature or rainfall and in some cases the quantity and quality of internally stored carbohydrates.

Growth may be either continuous from an apical bud at the top of the stem or the plants starts forming side shoots producing a bushy structure. In either case, environment will influence the switch-over from vegetative to reproductive growth. At that point, plants begin forming flowers where the reproductive organs swell either for wind or insect pollination. Once pollination is completed, the process of fertilisation leads to seed and fruit formation. Groups of plants which are grown for their seeds and fruit includes the legumes (peas and beans), sweet corn, cucurbits and courgettes, or crops where flowers are edible such as cauliflower and calabrese (green broccoli).

Fruit formation as in strawberry, raspberry and currants, is a response to fertilisation. Plants develop juicy structures which are attractive towards animals where colour, scent and possibly the presence of sugars are attractants. Animals eat the juicy fruit, digest them and then pass out the seeds through their digestive system where gastric juices may play a part in initiating seed germination.

Once plants have reproduced, a period of ageing (senescence) begins. This is not passive degradation as once thought. A suite of genes takes control of changes in leaf colour and the loss of green chlorophyll and development of vivid autumn colours (see Figure 3.68).

Figure 3.68 *Autumn colouration in Virginia creeper* (Parthenocissus quinquefolia)

These are a result from changing environmental factors, particularly the alteration from long days to shortening day length and declining temperatures. At this time, sap flow up the vascular tissues is reduced. Woody tissues in fruit crops such as blackcurrant ripen and start entering a dormant phase. The canes of raspberry dry out and die after fruiting. Strawberry plants lose their leaves and the exposed crowns (the leaves and buds of strawberry plants have a flattened rosette-like structure which is referred to as a crown) becoming resistant to the lowering of temperatures. Reducing temperatures encourages cold tolerance and subsequent dormancy in perennial plants. Annual plants lose their foliage and die off. Biennial crops like carrots and root brassicas may overwinter in the soil but are attractive to predators such as mice, slugs and wireworms because of their sugar content. These plants recommence growth as temperatures rise and there is free water available in the soil and the cycle of growth resumes in the spring. In biennial plants, this growth produces flowers and seed for future generations as seen in Figure 3.62, a flowering beetroot.

Achievements: have you understood and learnt?

At the end of the chapter, you should be able to:

1. Understand the processes which lead to the establishment of an independent free-living plant from seed and possible beneficial relationships with soil borne microbes.
2. Know how to prepare seed trays filled with appropriate seedling compost, distribute seed carefully and evenly on the compost and cover the seed with more compost.
3. Know that trays should be placed on a greenhouse bench and know how to water the compost, covering it with newspaper and glass sheets, subsequently ensuring the compost is moist.
4. Know that the emergence of seedlings requires monitoring, removing the newspaper and glass allowing the healthy growth of the seedlings.
5. Know how land is prepared for receiving the seedlings and transplants then carefully using the appropriate tools and erect supports for legumes as they develop.
6. Know that legume seeds may be sown directly into appropriately prepared land and establish appropriate protection and supports.
7. Understand that there are varying forms of plant and that they have been named according to their flower structure.
8. Describe the parts of a flower especially those related to sexual reproduction and the importance of chromosomes providing cells with information and instructions.
9. Describe the importance of asexual reproduction and the stages of mitosis in transferring genetic information into newly formed cells and resultant uniform growth.
10. Describe the importance of sexual reproduction as the vehicle for genetic variation and identifies the stages of meiosis by which male and female gametes are produced.
11. Describe the process of pollination and subsequent fertilisation and how fertilisation leads to the union of male and female gametes.
12. Understand the sources of seed for gardeners and the manner by which they should be stored.
13. Describe the differences between dicotyledonous and monocotyledonous plants and the importance of this differentiation for the culture of garden plants.
14. Understand how the form (morphology) and functioning (physiology) of plants are dependent on inherited characteristics passed from parents interacting with the environment and how this governs the cultural conditions required for the successful husbandry of plants of all types in the garden.

15. Describe the life cycles of plants, which vary in length depending on their genetic constitution and the environment in which they are grown.
16. Understand that each life cycle will commence with seed germination leading on to a phase of vegetative growth resulting in the formation of flowers which produce a new generation of seed.

Definition of terms: *Describe* is able to show why a scientific principle is important; *Understand* applies scientific principles into the context of gardening; *Know* carries through scientific principles in practical gardening activities.

Chapter 4
Growing small-seeded vegetables

Introduction

Growing crops from large seeds such as peas and beans is described in Chapter 3, and Chapter 2 deals with the use of asexually propagated plant parts such as potato tubers and onion sets. This chapter describes growing crops directly from small seeds and producing transplants for use in the vegetable garden. Care is required after sowing or transplanting which recognises and remedies emerging problems with weeds, pests, pathogens and nutrient shortages applies to all these crops. These hazards can cause major difficulties in ensuring that crops grow smoothly without competition or encountering damage from pests and diseases. Similarly, it is essential that disrupted growth and quality resulting from shortages of nutrition or water supply are avoided. This ensures that vegetables are succulent and well flavoured when eaten. The Chinese proverb that "the best manure is the farmer's boot" is very apt for most aspects of gardening, and in particular for crops grown from small seeds.

Garden practices

a. The range of small-seeded vegetables

Small-seeded vegetables include crops such as: beetroot, Brussels sprouts, cabbage, carrots, cauliflower, green broccoli (calabrese), leeks, lettuce, onion (spring and bulb) and parsnip. This group also includes crops such as courgettes and sweet corn which are cold-susceptible and initially need added care. Some of the Oriental small-seeded vegetables such as Chinese cabbage are also included; depending on the variety used they may be susceptible to cold damage or able to tolerate it.

Because all these crops are grown from small seeds, some only 2 mm in diameter, their supplies of energy are limited. Consequently, the gardener must provide optimal germination conditions, as explained in Chapter 1. Where germination is impeded, seedling growth becomes erratic, resulting in unrewarding plants. Erratic germination occurs most frequently either when the soil or compost is cold, which slows root development, or when there is too much water retained which encourages "damping-off" diseases (see Figure 2.41).

Cheap seed is an expensive mistake. Its germination will be poor and irregular and does not produce rewarding crops. Seed should be purchased from reliable sources and stored carefully as suggested in Chapter 3.

b. Direct sowing and transplanting

Small-seeded crops like carrots, parsnips and beetroot can be sown directly into the garden provided the soil is worked into a fine tilth and is adequately supplied with moisture. Robust crops of lettuce and brassicas are best obtained by raising transplants in a greenhouse or cold frame before moving them into the vegetable garden. Using transplants produces quicker maturity and ultimately higher-quality crops.

Alternatively, transplants may be bought either from a garden centre or some reputable national supplier. Some of these transplants will be ready for immediate planting into the garden, others which are smaller may first require protected culture in the greenhouse. Plants purchased in a garden centre for either option should be closely inspected before purchase. Ensure that these plants are standing erectly and robustly in their containers with no signs of wilting or yellowing. The presence of these symptoms indicates a shortage of nutrients in the compost, which most probably results from having been held for too long by the garden centre. Shop displays should be lit with natural light and with a good air flow but not placed in a windy corner. Avoid plants that have stood out in the wind and rain without protection. Although the plants may not immediately show signs of distress from such conditions, they will have suffered stresses which ultimately cause unthrifty growth. Garden centres and other shops selling transplants should have a member of staff who is qualified in their care and can knowledgeably advise customers.

Figure 4.1 *Cell tray for sowing lettuce seed*

c. Growing transplants under protection

Small seeds should be germinated carefully and individually. This is best achieved by using modular or cellular trays in which each plant has an allocated space and volume of compost. Modular or cell trays can be purchased from garden centres, shops or mail-order suppliers (see Figure 4.1).

Fill each module or cell in the tray with seeding compost to within 10 mm of the top. This can be gently firmed by tapping the rim or side of the tray settling the compost. As with pea and bean seeds in Chapter 3, open the seed packet and place some seed on to a plastic or clay saucer (see Figures 4.2 and 4.3).

This makes subsequent handling easier, more convenient and safer. Using a short length of bamboo cane or a dibber made for this purpose make a shallow hole in the compost. Picking up individual seeds should be done carefully, always being conscious that they are dormant but living entities and hence can be damaged by rough handling. Manually place a small sample of seed in the palm of one hand and with the fingers of the other hand select single seeds placing one seed in each hole in the centre of the selected module or cell tray (see Figure 4.4).

It is possible to purchase devices for sowing small seeds which will drop them singly into the holes in the compost, or alternatively they can be gently selected using a pair of forceps. Repeat the process of seed sowing until all the modules or cells have been filled. Using a garden sieve, shake a sufficient but small amount of fresh compost over the module tray, filling up the holes. Firm the compost by tapping gently on the side of the tray.

Place each tray on the greenhouse bench and water from a watering can using a very fine rose (see Figures 4.5 and 4.6).

Label each tray with the seed type and date of sowing. There is a convention that labels should be placed in the top right-hand corner of seed trays or rows of pots. Cover each tray with a pane of glass, with some old newspaper placed on top as a form of insulation (see Figures 4.7 and 4.8).

Figure 4.2 *Lettuce seed packet*

Figure 4.3 *Lettuce seed*

Figure 4.4 *Lettuce seed sown and ready for "dibbing-in"*

Figure 4.5 *Sown cell trays of lettuce seed placed on greenhouse staging*

Figure 4.6 *Water cell trays sown with lettuce seed*

Figure 4.7 *Cover cell trays sown with lettuce seed with sheets of glass*

Figure 4.8 *Sown trays of lettuce seed covered with sheets of glass, with old newspaper*

Inspect the trays at daily intervals, ensuring that the compost remains damp but does not become waterlogged, turn the pane of glass over, thereby removing condensation. As soon as there are any signs of emerging seedling shoots, remove the newspaper but leave the glass in place for a short time. This helps retain moisture in the compost and maintain favourable conditions for germination. As soon as the seedling shoots approach the glass, it should be removed quickly, otherwise they will become distorted. Keep the seedlings regularly watered so that the compost is moist but not saturated. Similar procedures may be used for brassica seed (see Figure 4.9).

The modular or cell trays selected should be of sufficient size that the plants can make adequate growth before they are transplanted (see Figure 4.10). Achieving this requires that the greenhouse environment provides an effective airflow produced by manipulating the ventilation. All greenhouses are fitted with ventilators. These should be opened early in the morning and closed down last thing at night, especially if there are warnings of frost. During spells of warm weather in spring and summer, it may be wise to leave the ventilator slightly open at night so that there is no overheating early the following day. In winter, the main purpose of ventilation is removing excessive atmospheric moisture. As a rule of thumb, ventilators should be capable of providing an air-change every 20 minutes in cool conditions and more frequently when temperatures rise.

Lettuce and cabbage (brassica) transplants may be purchased from garden centres in place of those raised by the gardener. These have the advantage of avoiding the necessity for growing from seed under protection (see Figures 4.11 and 4.12).

d. Transplanting into the garden

When the lettuce or cabbage seedlings are of sufficient size, normally showing 3 to 4 true leaves, they may be transplanted into the open (see Figures 4.13 and 4.14). At this growth stage, they may be handled easily. Prepare an area of open ground for transplanting by lightly raking it into a fine tilth. Sprinkle "GrowMore" general fertiliser at 30 g per square metre over the surface

Figure 4.9 *Cabbage seed for sowing in cell trays*

Figure 4.10 *Lettuce plants in a cell tray ready for transplanting*

Figure 4.11 *Commercially grown lettuce plants in cell trays ready for transplanting*

Figure 4.13 *Lettuce transplant showing a healthy root system ready for transplanting*

Figure 4.12 *Display of commercially grown brassica (cabbage) plants ready for transplanting*

Figure 4.14 *Cabbage plant showing a healthy root system ready for transplanting*

and rake it into the top 5 cm of soil. Using a garden line, identify a straight row for planting. Take the module or cell tray and gently squeeze each tube around its sides. If there is a large enough hole at the base, it helps to push the seedling out of the module by placing a finger or a short bamboo cane in that hole. This releases the transplant module from its tray without damage. When seedlings have produced a vigorous root system, they will fully fill the module. That binds the compost together, ensuring that the module can be transplanted in its entirety.

With a trowel, take out a shallow hole which has just sufficient volume and will accommodate the module and its entire root system without cramping. Press the soil back round the module, refilling the hole so that the root system is fully covered. The transplant should be placed in the soil such that the neck of the plant is level with the soil surface. Once all the plants are placed in their positions for cropping, then water well using a can fitted with a coarse rose, and sprinkle some slug repellent round the young plants. This process should be repeated until all the lettuce, cabbage or green broccoli transplants have been relocated into the soil or raised bed compost. There is close resemblance between planting lettuce and cabbage transplants and the techniques used for beans, as described in Chapter 3. Lettuce are very useful salad vegetables for smaller gardens as they mature in 4 to 6 weeks after transplanting (see Figure 4.15).

Raising and planting batches of one or two dozen plants at a time will provide a sequence of crops from early June through the summer. Planting larger numbers runs the risk that they will become over-mature and "bolt", producing a flower stalk. Bolted lettuce has little or no use in the kitchen. Germinating lettuce seed through the summer may become more difficult. Lettuce seed is thermo-dormant; it does not germinate at temperatures above 12 to 15 °C. Consequently, for late summer and autumn crops, purchasing plants from a garden centre or mail-order supplier may be the gardener's best policy.

Figure 4.15 *Maturing lettuce plant in the garden*

e. Directly sowing crops

Sowing beetroot

Beetroot, carrot and parsnip seed is sown directly into the ground using some of the practices already described in Chapters 2 and 3. Here, beetroot is used as an example, each seed is a complex cluster of individuals (see Figure 4.16).

With a rake, work down an area of soil to a fine tilth adding "GrowMore" general fertiliser at 30 g per square metre (see Figures 4.17, 4.18, 4.19 and 4.20).

Identify a straight row with the garden line (see Figure 4.21) and draw out a drill with the hoe blade placed at right angles to the soil surface alongside the string (see Figure 4.22).

Figure 4.16 *Beetroot seed ready for direct sowing in the garden (note that these are seed clusters, as shown also in Figure 3.62)*

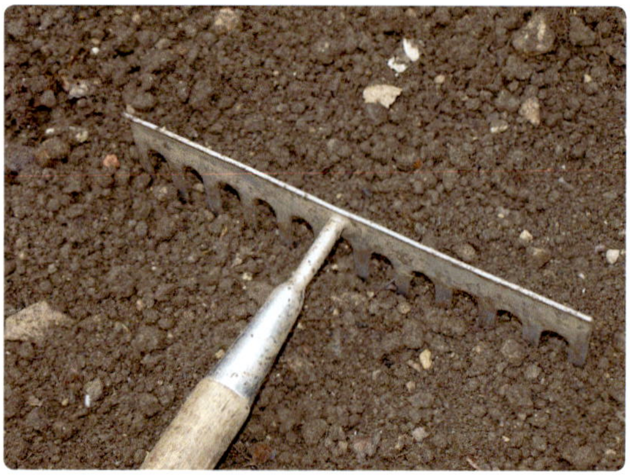

Figure 4.17 *Raking soil to a fine tilth*

Figure 4.18 *Comparison of fine tilth (lower half of the picture) and a coarse tilth (upper part of the picture)*

Lightly moisten the bottom of the drill from the watering can fitted with a fine rose. Be careful not to over-moisten the row, as this can make it waterlogged and then the seed will germinate erratically.

Open the seed packet and pour a small quantity of seed into the palm of the left hand. With the right-hand fingers, pick up a small quantity of seed and sprinkle it along the drill (reverse holding pattern for left-handed people) (see Figure 4.23).

Figure 4.19 *Scatter fertiliser granules evenly over the fine tilth*

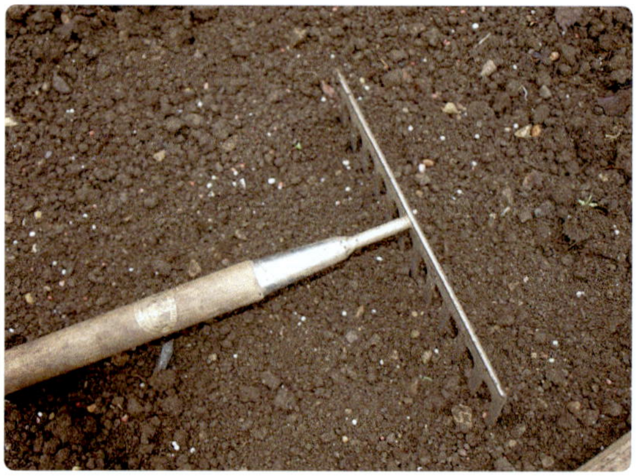

Figure 4.20 *Raking in the fertiliser granules*

Gently fill in the drill with fine soil using the hoe blade and firm the drill with the back of the hoe blade (see Figures 4.24 and 4.25).

Water the drill from the can fitted with a coarse rose.

Figure 4.21 *Place a garden line used to guide a straight row (drill)*

Figure 4.22 *Take out a drill for the beetroot seed with the edge of a hoe*

Mark the position of the drill with short bamboo canes placed at either end of the row. Sprinkle slug repellent over the drill.

Where appropriate, place wire netting, plastic cages or polythene or glass cloches

Figure 4.23 *Space the beetroot seed sown in the drill*

Figure 4.24 *Covering the seed with soil using the blade of a hoe*

Figure 4.25 *Completed and covered drill*

Figure 4.26 *Water the drill when completed*

Figure 4.27 *Cover the row with glass or plastic cloches to accelerate germination*

Figure 4.28 *Beetroot seedlings emerging*

Figure 4.29 *Beetroot seedlings thinned (spaced) out, allowing the remaining plants ample room for growth*

over the transplants and seed drills as protection from vermin such as wood pigeons and as a means of accelerating growth (see Figure 4.27).

As the seedlings germinate and emerge, or the transplants become established, weed competition may become a problem. It is very important that both seedlings and transplants are not subjected to competition for light, nutrients and water by weeds or from the presence of too many plants (see Figure 4.28).

This means that the density of seedlings must be reduced by the process of "thinning out" (see Figure 4.29).

Minimising competition within the crop is as important as excluding weed growth. Populations of directly sown seedlings will require thinning out once they are sufficiently large for handling. This gives the remaining plants sufficient space in and between rows for growth.

It is essential for direct-sown crops such as beetroot, carrots or parsnips.

Thinning out is achieved by gently removing excess seedlings, leaving one plant every 3 to 5 cm. Retain the most robust and vigorous plants and remove any which are malformed or small relative to the rest of the population in the row. Irrespective of the weather conditions, always water those seedlings retained after thinning out, this re-settles the soil back around their roots and moistens it. This operation can be avoided by using seed that is pelleted or sold embedded in soluble tape.

At the young growth stage, seedlings and transplants can also be badly retarded by weed competition. Hand weeding is a laborious but essential process (see Figure 4.30).

Residual beetroot plants will make robust growth following thinning out and weed removal (see Figure 4.31).

The plants should also be monitored for signs of pest and disease development (see Underpinning knowledge 4.3). Regular attention, weed removal and the eradication of pests and diseases will provide the gardener with good crops of beetroot (see Figure 4.32), carrots (see Figure 4.33) and cabbage (see Figure 4.34).

Figure 4.31 *Vigorously growing beetroot plants*

Figure 4.32 *Mature and harvested beetroot*

Figure 4.30 *Removing weed seedlings alongside beetroot seedlings with an onion hoe*

Figure 4.33 *Mature and harvested carrots*

Figure 4.34 *Mature 'Hispi'-type spring cabbage ready for harvest*

f. Cold-sensitive small seeds

Courgettes and sweet corn originate from warm temperate and semi-tropical regions in Asia and South America, respectively. Consequently, they are susceptible to low-temperature damage. That means that temperatures as low as +5 °C will cause damage. As a precaution, the gardener should either buy transplants from a garden centre when all likelihood of frost is past or germinate plants from seed in the greenhouse.

Sweet corn is a good example of the manner by which varieties have been improved over years by plant breeders. The gardener will find a range of varieties offering kernels (seeds on the cobs) with different degrees of texture and sweetness. These are broadly classified as: "normal sweet", having a mild flavour but which can be lost after picking or if left on the plant too long after maturity when the tassles (part of the female flowers) turn brown; "sugar enhanced", which are sweeter, with a crisp and creamy texture and flavour; "supersweet", which retain sweetness longer but are not creamy; "extra tender or enhanced", where the kernels are even more tender and sweetness lasts longer as a result of considerable hybridisation.

The seeds of both these crops are quite sizeable and less fitted for propagation in modular or cellular trays. Using 10 cm diameter (at the top rim) plastic pots, possibly collected and cleaned after the purchase of bedding plants, fill with seedling compost to within 1.5 cm of the top. Firm the compost using the bottom of a suitably sized pot. Holes 2 cm deep are made with a piece of bamboo cane as a dibber in the compost filling each pot. Having opened the seed packet and placed all the seed in a plastic or clay saucer select individual seeds and place one in each hole (see Figures 4.35 and 4.36).

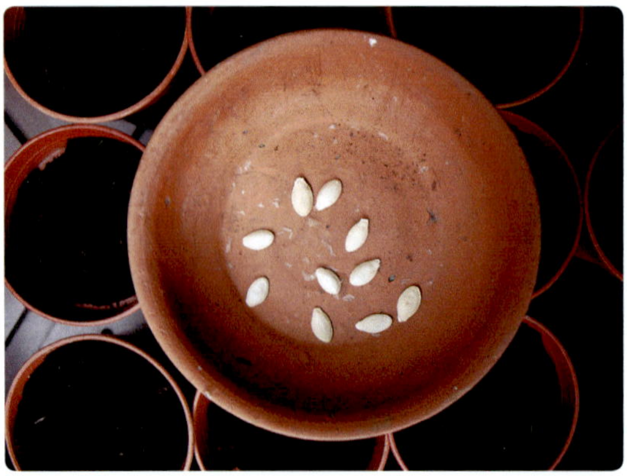

Figure 4.35 *Courgette seed*

Figure 4.36 *Courgette seed sown individually in pots of compost*

It is probably best to keep courgettes separated individually one plant per pot as they grow very rapidly (see Figure 4.37).

Germination of sweet corn is slightly slower and 2 seeds may be placed in each pot. When sowing is completed, cover the surface of the compost with a 0.5 cm layer of compost from the garden sieve. Place the pots on the greenhouse benching or in the cold frame, and care for them in a similar manner to all other seedlings (see Chapter 3). Both these crops may germinate and produce shoot growth within 7 to 10 days if conditions are warm enough (see Figure 4.38).

These plants will grow rapidly in their pots and should be regularly monitored for signs of water shortage. In warm and or sunny weather, these plants will require watering from a can fitted with a fine rose on a daily basis, and possibly more frequently. Once the courgettes have developed one true leaf and the sweet corn plants are 6 or 8 cm tall, they can be transplanted in a similar manner to other vegetables (see Chapters 2 and 3).

Courgettes will form very large spreading plants when they reach full cropping size and sweet corn grows erectly to 2.5 m tall if well fertilised (see Figure 4.39). Both crops will require sufficient garden space for their development and cropping. Courgettes have gross requirements for water when fully fruiting. As soon as courgette fruit are 6 to 8 cm long, they should be harvested (see Figure 4.40). Allowing more growth reduces future productivity, as the plant will divert its resources into seed formation inside the fruit, which ultimately swells becoming a marrow (see Figure 3.61).

Figure 4.37 *Courgette seedlings*

Figure 4.38 *Sweet corn seedlings*

Figure 4.39 *Courgette plants in flower and with fruit*

Figure 4.40 *Mature courgette fruits harvested*

Sweet corn plants may require support as they reach maturity, especially if the garden is in a windy location. It is wise to place a stout post at either end of the row and wire or string along the row, perhaps with some large intermediary bamboo canes providing support as the sweet corn matures (see Figure 4.41).

If necessary, the sweet corn stems can be tied with 5-ply garden string to the wire for additional support. The volume and weight of leaves and stems produced by these plants is very substantial, they also produce very substantial penetrating roots (see Figure 4.42).

Figure 4.41 *Sweet corn male (upper) and female (lower) flowers and an adjacent row of courgettes*

Figure 4.42 *Vigorous root systems of mature sweet corn plants*

Sweet corn produces separate male and female flowers on the same plant. The male flowers develop at the tops of stems with very prominent anthers. When these flowers mature, pollen is shed and drifts downwards onto the developing female cob, which has at that stage translucent to white styles and stigmas in tassels protruding beyond the flower (see Figure 4.43).

Once pollination takes place, the tassels turn brown and the cob of seeds develops wrapped in a sheath of green leaves (see Figures 4.44 and 4.45).

Cobs should be harvested once all the seeds along the entire surface have turned a rich yellow colour when flavour is highest and at its most enjoyable.

g. Traditional winter vegetables: leeks

Leeks are a valuable winter vegetable although there are now varieties which extend the growing season from late summer through to the

Figure 4.43 *Sweet corn plants with developing cobs*

Figure 4.44 *Sweet corn cobs with seeds*

Figure 4.45 *Pair of mature sweet corn cobs*

end of the following spring if a succession of supply is required. The culture of these different varieties is similar. But the late-season varieties require longer growing periods relative to early-season types. Leeks are best grown from seed germinated in modular or cellular trays like other vegetables. They should be allowed to produce 25 to 30 cm of leaf growth before transplanting in the late summer. Late planting is a useful characteristic since it means that a particular area of the vegetable garden can produce more than one crop per year. Leeks are also useful because they form large and vigorous root systems, which helps create a well-aerated and healthy soil. Leek roots exude several natural chemicals which are capable of combating some soil-borne diseases, so they also have a valuable role as bio-controlling plants (see Underpinning knowledge 2.3). The growth of leek root systems requires stimulation at transplanting, and this is achieved by cutting back the green parts of the leaves (see Figures 4.46 and 4.47).

This unbalances growth patterns between the green and the lower blanched areas of leaf, with the effect of encouraging the development of the latter.

That is achieved by cutting off half of the leaf growth. Cut the leaf growth down from 25 to 30 cm to 10 to 15 cm with a sharp garden knife or a pair of shears. Planting requires similar land preparation as described for other transplants using a garden line and dibber (see Figure 4.48).

Once the garden line is in place, make a hole with a dibber which is 10 cm deep and place a leek plant in it (see Figures 4.49 and 4.50).

But do not refill the hole, simply water well so that the soil gently fills back around

Figure 4.46 *Leek transplants with vigorous root systems*

Figure 4.48 *Dibber for planting leek transplants*

Figure 4.47 *Leek transplants with their foliage reduced*

Figure 4.49 *Dibbing-in (making a suitably deep hole) for leek transplants*

Figure 4.50 *Leek transplants placed in their stations*

Figure 4.51 *Watering in leek transplants*

the module and the base of the plant (see Figures 4.51 and 4.52).

This practice encourages blanching of the leek stem as it grows and expands. The length of blanched stem may be increased by mounding up earth around the developing leek foliage as it grows. But leave sufficient top growth so that photosynthesis continues. Leeks are gross feeders requiring additional top dressings of fertiliser possibly applied as liquid feed during the growing season, these liquid feeds should have a high nitrogen content. The plants will continue growing in winter and that accelerates in the early spring as soil temperatures rise producing very acceptable vegetables (see Figure 4.53).

This type of growth pattern reflects the origins of leeks from very cool temperate areas. Some hybrid leek varieties grow and blanch without moulding up the soil. These can be grown year-round under protection using a walk-in polythene tunnel. Plants can be purchased as transplants for this purpose from some mail-order sources or garden centres.

Figure 4.52 *Leek transplants with soil settled around them*

Figure 4.53 *Mature leek*

h. General husbandry care

All vegetable crops grown either from directly sown seed or put out as transplants require similar general husbandry care. Essentially this consists of:

- Preventing weed competition because this reduces the nutrients available for crops, limits regular and even growth by excluding light and offers havens for pests and diseases.
- Protecting crops from predation by pigeons, these both eat crops and spoil them with their excrement; netting covers are an essential precaution especially during winter and early spring when other food sources are scarce.
- Adding more granular fertiliser or use liquid preparations as top-dressings. These latter have standard nutrient formulations or organically supportive feeds such as seaweed extracts or compost teas.

A summary of times for sowing, planting and transplanting vegetables together with suggestions for seed sowing depths, distances between rows and plants is given in Table 4.1.

Also note that the monthly timings are approximate and will vary between seasons depending on weather conditions and geographically. Gardeners living in the southern counties of England can work between 2 and 4 weeks earlier than those who are further north, and in Scotland these differences may extend to 6 or more weeks. But again, this will depend on season and to a considerable extent the location of a garden.

Underpinning knowledge

4.1 Shapes, sizes and fitting together

Gardening is about providing environments in which plant growth is controlled and manipulated, this delivers satisfaction and enjoyment gained from home-grown vegetables, fruit and flowers. Achieving satisfactory crop growth requires an understanding of the shapes and sizes of plants and fitting them into the patterns and timings which the gardener designs. This means moulding the genetically determined architecture of plants into the geometry required for successful crops.

In nature, plant shape results from the genetic information inherited from both parents interacting with the environment in which they are growing (see Underpinning knowledge 3.1). In the garden, husbandry practices artificially and intentionally control this balance. The gardener's knowledge and skill determine how plants grow. For millennia, gardeners have shaped plants aiming at gaining added beauty, quality and yield; now they are aided by plant breeders who alter the genetic architecture, specifically enhancing characteristics suitable for gardening.

Plant structure

Botanically, plants consist of roots, shoots, leaves and flowers. Although, in some plants, there is a distinct zone above the root known as the hypocotyl above which primary leaves, cotyledons, develop (see Figure 4.54).

Table 4.1 *Calendar for sowing, transplanting and harvesting vegetables**

Crop	Sow covered	Sow outside	Plant outside	Transplant	Harvest	Sowing depth (cm)	Space between rows (cm)	Space between plants (cm)
Broad beans	Jan–Feb	Feb–April and Oct–Nov	April–May		June–Aug	5	45	20
French beans	Feb–March	May–June	May		July–Oct	5	35–45	10–15
Runner beans	April–May	May–June	May–June		July–Oct	5	45–60	20–25
Beetroot	March	April–Aug			June–Nov	2.5	30	10
Broccoli/ calabrese	March	March–July	May	May–July	Jan–April and July–Dec	0.5	45.60	45
Brussels sprouts	Feb–April	March–April	May	May–June	Jan–March and Sept–Dec	1.25	60–75	60–75
Cabbage – spring		May–Aug		Sept–Oct	March–Dec	1.25	45–60	30
Cabbage – summer and autumn	Feb–March	March–May	May–June	May–June	June–Nov	1.25	30–45	30–45
Cabbage – winter	March	April–May	May–June	June–July	Jan–May and Nov–Dec	1.25	45	45
Carrot	Feb	March–Aug			Jan and June–Dec	1.25	30	2.5–5.0
Cauliflower – summer	Jan–Feb	March–May	March–May	May–June	May–Nov	1.0	60	60
Cauliflower – winter	April	May–June	May–June	June	Feb–May	1.0	60–75	60
Celery	Feb–April		May–June		Aug–Nov	0.5	20–25	20–25
Celeriac	Feb–April		May–June		Jan–March and Sept–Dec	0.5	45	30
Chinese cabbage		April–May and July–Aug			June–Oct	1.0	35	25
Courgette	April–May	May–June	June		June–Nov	2.5	90	90
Florence fennel		May–July			July–Oct	1.0	45	20
Kale	March	April–July	May	June–July	Jan–May and Nov–Dec	2.5	45–60	45

(Continued)

Table 4.1 *(Continued)*

Crop	Sow covered	Sow outside	Plant outside	Transplant	Harvest	Sowing depth (cm)	Space between rows (cm)	Space between plants (cm)
Kohl rabi		April–July			June–Nov	0.5	30	15
Leeks	Jan–Feb	March–April	May–June	June–July	Jan–March and Sept–Dec	1.0	30	15
Lettuce	Feb	March–Sept	March–April	April–Sept	March–Sept	1.0	30	15–25
Onion – bulb	Jan–Feb	March–April	April–May		Aug–Sept	1.5	30	10
Parsnip		Feb–May			Jan–April and Sept–Dec	2.5	30	15
Peas	Feb–April	March–June	April–May and Oct–Nov		June–Oct	5.0	60–90	5–10
Radish	Feb–March	March–Sept			April–Oct	1.0	15–20	2.5
Rocket	March–April	March–Aug			May–Nov	1.0	35–45	15–20
Spinach	March–April	March–Sept			April–Oct	2.5	30	7.5–15
Onions – spring	March–April	March–Aug			July–Oct	2.5	10	2.5
Swede		May–June			Jan–Feb and Sept–Dec	2.5	30	20
Sweet corn	March–May	May	June	June	Aug–Sept	2.5	45–60	45–60
Swiss chard	March–April	April–Aug			April–Oct	2.5	35–45	20–30
Turnip	Feb–March	April–Aug			April–Dec	2.5	30	15–25

Source: Adapted from the *Vegetable Seed Growing Guide* published by Marshalls Seeds Ltd, Cambridgeshire.

Note: *There is limited difference between planting outside and transplanting, the former refers to plants brought into the garden and the latter to plants raised outside and transferred within the garden.

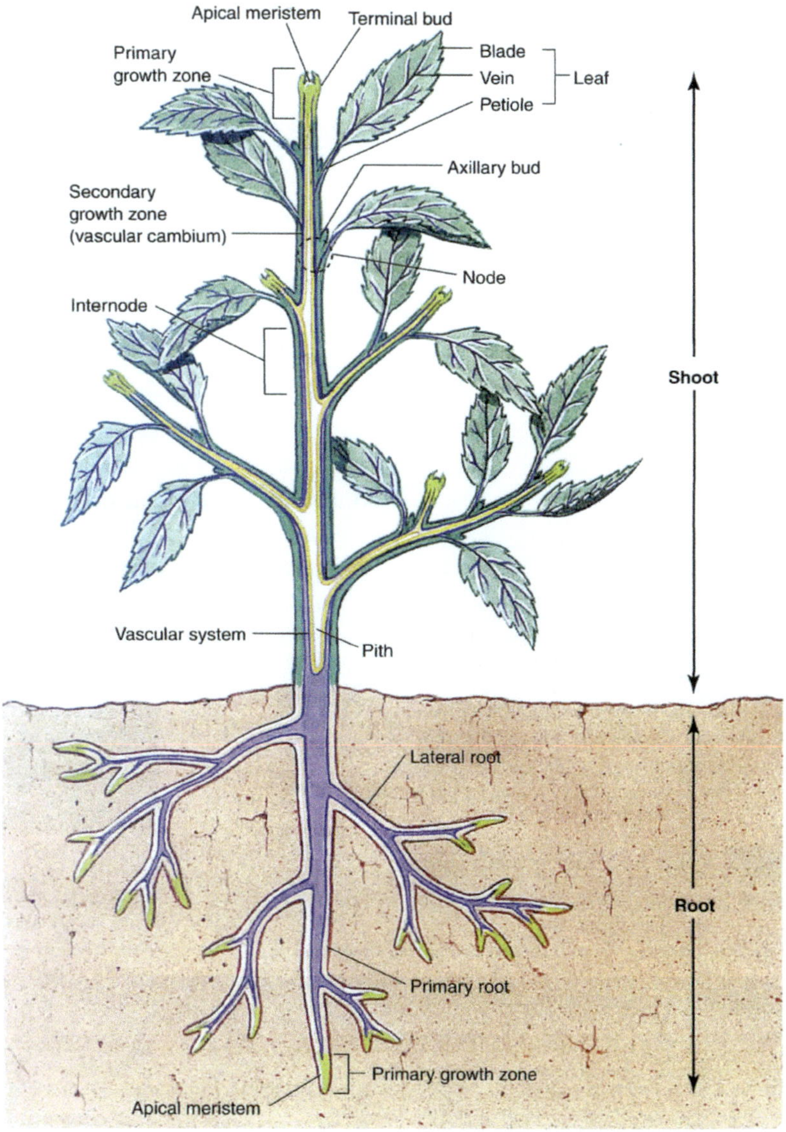

Figure 4.54 *Basic plant structure*

Connecting these is a series of specialised tubes running between roots in the soil and shoots, leaves and flowers in the aerial environment, known collectively as the vascular tissue (see Chapter 1). At either ends of roots and shoots are layers of unspecialised but vigorously mitotically dividing cells, the meristems (see Underpinning knowledge 3.1 and 3.2). This multiplication of cells results in the lengthening growth along the whole plant. As described in Underpinning knowledge 3.1, each living cell contains a nucleus within which are the chromosomes carrying the genes inherited from the parents and bounded by a nuclear membrane. The remainder of the cell is filled with cytoplasm in which are a series of structures (organelles), each having very specialised functions. All cells contain mitochondria which are the powerhouses where energy transfers take place and ribosomes where coded information from the nucleus delivered by messenger RNA is transcribed and translated into instructions controlling cell processes. Plant cells exposed to light contain chloroplasts, the centres for photosynthesis (see Underpinning knowledge 1.1 and especially Figure 1.9).

Figure 4.55 *Annual rings in a tree section*

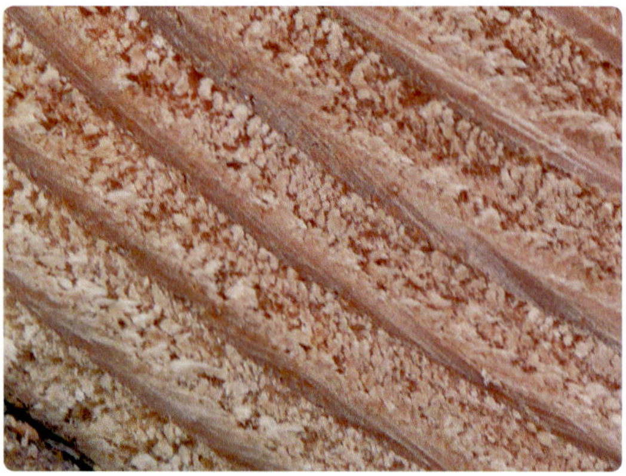

Figure 4.56 *Close-up of annual rings showing the variation in cell size*

Minerals, water and the products of photosynthesis are transported about plants in the tubes connecting the roots and stem apex known collectively as the vascular tissue. The outer vascular layers are composed of living cells, the phloem, which are conduits moving the products of photosynthesis out from leaves and stems down towards the roots and also distributing resources upwards. At the centre of vascular tissue are layers of dead cells (xylem) conducting water and nutrients up from the roots into stems and leaves. Between the phloem and xylem are more layers of reproductive meristems producing new vascular tissues and resulting in the lateral expansion of plant girth in dicotyledonous plants especially trees and shrubs (see Underpinning knowledge 1.2 and see Figures 5.51 and 5.52).

Stems

Stems are the conduits through which water and nutrients are transported to the leaves and the products of photosynthesis distributed downwards, ultimately reaching the roots or upwards to sinks such as flowers or fruit. They produce leaves from buds initiated in asexually reproductive cells within the stem and which may then form extension growth forming lateral shoots, eventually forming branches. This process is also known as vegetative reproduction. The cylindrical nature of stems provides physical flexibility capable of withstanding stresses and strains caused by wind and rain. Stems vary enormously in structure and functions from those which seek support from other plants for example, runner beans compared with those which form large trunks as in trees. Stem expansion in woody trees and shrubs living for several years or more, results in annual growth rings of cells (see Figures 4.55 and 4.56).

The size of annual rings reflects the environment in which plants have grown. When the weather favours plant growth, large cells form in the annual ring; harsh conditions result in thinner rings with smaller or fewer cells.

Surrounding the vascular tissue in green stems are layers of thin-walled cells; these contain chloroplasts the numbers of which increase towards the outside. Wrapped round the entire structure is a single layer of cells with strengthened walls, the epidermis. A biologically complex and porous waxy layer, the cuticle, provides protection from rain and deters herbivorous predators and pathogens. As stems age in trees and shrubs, the outer layers become an inert bark again, with protective functions against herbivores, pathogens and, in some trees, such as the Californian Giant Redwood (*Sequoiadendron giganteum*), or Australian *Eucalyptus* spp., protection from fire.

Roots

Roots are the entry portals for water and nutrients taken in from the soil or other substrates. Increasingly, research is showing that roots are capable of communicating by chemical messaging molecules with each other and with microbial communities in the soil. There is a thin and biologically very active soil layer, the rhizosphere, which surrounds roots. Within this zone populations of benign microbes provide protection from pathogenic types and in return utilise root exudates containing complex carbon compounds which bacteria and fungi cannot produce. Physically, roots provide anchorage for the aerial organs of plants (see Chapter 1), they may penetrate the soil very deeply with tap roots or more superficially with fibrous systems. The end of each root is strengthened with a protective cap, enabling uninjured extension growth between the abrasive soil particles. Even with this protective cap, roots penetrate more easily and with less damage through soils which are friable and have a tilth of small crumbs as opposed to those composed of large clods or with an airless compacted layer (see Chapter 2). Behind the cap are very large numbers of delicate single-celled root hairs. These provide a large surface area capable of rapidly absorbing water, nutrients and oxygen from the soil. Root hairs are very transitory structures, maturing and being replaced quickly in daily cycles as the root pushes further forward through the soil.

Root shape varies largely between annual and perennial plants. Annuals generally have large fibrous root systems which spread laterally and quickly absorb resources from the soil. The alternative structure is a single tap root growing down with relatively few side roots. Tap roots are typical of trees and shrubs grown from seed. Container-grown and containerised plants propagated from cuttings or those sold as bare-rooted transplants purchased from nurseries and garden centres will probably lack a tap-rooted structure (see Figure 8.18). When there are periods of water shortage, fibrous-rooted plants wilt more quickly compared with those that are tap-rooted and capable of seeking water at greater depths in the soil.

Leaves

Leaves are the main centres for the manufacture and consumption of complex carbon compounds through the processes of photosynthesis and respiration (see Underpinning knowledge 1.1 and 1.2 and Figure 1.9). Water is translocated from the roots through stems, into the leaves

Figure 4.57 *Stems, leaves, shoots and buds illustrated by a herbaceous* Salvia *spp. plant*

in the transpiration stream which results from a suction pressure generated by evaporation of water through the stomata (see Underpinning knowledge 1.3). Leaf form varies enormously as a result of evolutionary selection fitting them for compatibility with the prevailing natural environment. For example, the evergreen leaves of most conifers have cylindrical needle-like shapes which successfully survive excessive cold, heat or wind in the northern and southern cool temperate regions from which they originated. By contrast, tropical plants produce large flattened leaves capable of withstanding very humid atmospheres.

Leaves usually consist of a network of conducting vascular tissues surrounding these are layers of thin-walled cells containing chloroplasts. The upper surfaces of most leaves are covered in thick protective cuticles. On the under surface are openings known as stomata through which gases and water vapour are exchanged with the surrounding atmosphere. Oxygen needed for respiration is mainly taken in at night, while carbon dioxide is used during periods of daylight for photosynthesis. Photosynthesis (see Chapter 1) releases oxygen which plants can utilise in daylight. Water vapour is transpired out from the stomata. These structures are illustrated in Figure 4.57.

Buds

Buds are asexually and sexually reproductive centres. Asexually, they provide vegetative extension growth which forms the characteristic shapes of plants. Sexually, buds develop into flowers and fruit. The two forms of reproduction (see Underpinning knowledge 3.1) enable the survival of plants through adverse environmental conditions or by producing new generations capable of responding by evolutionary change. Buds develop in the axils between leaves and stems forming vegetative shoots and eventually branches. Alternatively, when correctly stimulated the buds become sexually active forming flowers and eventually dry or succulent fruits containing seeds. Conversion from vegetative to reproductive states requires stimuli from the environment, such as changing day length, periods of rain or significant changes in temperature, plus the genetic instructions which organise these changes coming from the nuclear DNA and converted into action via RNA in the ribosomes (see Underpinning knowledge 7.1). The nature of the stimuli depends largely on climatic conditions in the geographical zone where particular plants evolved. Changes are regulated by internal formation and release of growth-regulating hormones, leading towards the onset of cell division, elongation and the alteration of functions. Ultimately, that initiates the processes of fruit and seed formation accompanied by ripening, which are particularly important for gardeners. Horticultural scientists now recognise that each change in structure and function is controlled by genes inherited from the parents, information from which is transferred from the nucleus to the ribosomes in the cytoplasm from where growth and reproduction are organised.

Plant habit

Habit is a term which describes the physical appearance of plants, which may change quite radically as they grow, flower and reproduce. Planning garden layout requires knowledge of plant structure and functioning. For example, annual vegetable or herbaceous plants may grow solely from the tips of stems or branches producing either columnar or bushy habits. The forms of growth are known as indeterminate or determinate; in the former, the plant continues growing from the apical bud at the top of the stem, and in the latter growth develops from lateral shoots on the stem, giving a bushy structure.

Differences in leaf shape influence, for example, planting densities. Onions, which have long thin tubular leaves, and can be planted more densely than plants such as cabbages or potatoes, which have spreading habits. Some plants such as peas and runner beans are climbers requiring structural supports. The extent of the climbing habit determines the size and shape of supports that gardeners require for their crops. Crops maturing early in the season tend to be shorter than those which are maincrop varieties. Knowledge of these variations is crucial for planning throughout the garden.

Size and population

Understanding the rate at which plants grow and their ultimate size regulates the amount of land earmarked for particular crops and even the amount of weed control required. Crops such as potatoes or courgettes have large rapidly spreading habits. They grow quickly forming dense canopies of leaves and smother weeds. Lettuce will eventually form large canopies but because of its slower growth rates demands weeding in the early stages (see Figures 4.58 and 4.59).

Figure 4.58 *Rampant annual weeds*

Figure 4.59 *Maturing lettuce surrounded by annual weeds*

The use of transplants grown initially under protection is a valuable technique for lettuce because they can then form canopies more quickly which suppress weed growth. Controlling weed competition brings the bonus of helping avoid infection with some aphid-borne virus diseases harboured by for example chickweed (*Stellaria media*).

Beetroot, carrot and parsnip crops are best grown by directly seeding into the garden. Although they can smother weed competition as mature plants it must be controlled in the early growth stages otherwise yield and quality will be severely reduced. These crops will also require thinning, those left un-thinned will produce smaller roots in which thick cell walls have developed and products which are hardly worth harvesting. There is an additional danger in that plants grown at too high densities are often more susceptible to pests and diseases and more prone to physiological stresses or nutrient deficiencies.

Understanding the relationship between size and effectiveness is equally important in the ornamental garden. Where shrubs and herbaceous plants are grown together knowledge of their eventual height, girth, colour and capacities for expansion is essential. Understanding these characteristics permits the development of planting schemes where specimens complement each other and prevents a particularly rapid growing and dominant type from colonising the entire border (see Underpinning knowledge 7.2, 7.3 and 7.4). In planting schemes, use can be made of the changing structure and appearance of plant foliage and flowers as they develop and age. An extreme example of this is the use of some dwarf grasses such as forms of *Miscanthus* which, even when senescent, can form useful foils for the colourful stems of shrubs such as *Cornus* in winter.

Bulbs can make considerable contributions for border planting for example, especially early or late in the seasons but because of their habit planting should be planned with care. They have upright and tubular foliage and do not compete well with some of the early vigorous broad-leaved plants (see Underpinning knowledge 6.2).

There are similar requirements for cautious planting in the fruit garden. Raspberry varieties vary substantially in the volume and vigour of their early-season shoot growth. The Scottish-bred types which have names prefixed with "Glen" can be very vigorous especially when grown in southern England. Consequently, cane growth requires restriction before fruiting, otherwise picking is made more difficult and soil nutrient reserves become exhausted more quickly by supplying energy into unwanted foliage. Bush fruits such as currants form large plants within five or six years of establishment; consequently, gardeners need relatively few plants and should allow sufficient space for them as they mature. Strawberries can become rampant after fruiting producing copious asexually reproductive runners (see Chapter 5). These require thinning and removal of about 95 per cent of their growth, otherwise the whole bed becomes a mass of spindly plants, diminishing future yields. Numerous other rampant berry crops are now on the market, such as variants of the blackberry. These, once established in the garden, can form huge plants wholly unsuited for small urban plots (see Chapter 5).

4.2 Ways and means

Evolving effective ways and means for using resources lets plants survive in nature, the gardener artificially redirects these processes into successful cropping. Plants take resources from the air and soil. The air supplies energy as sunlight, plus the gases carbon dioxide, oxygen and nitrogen. Soil or other substrates supply water and nutrients (see Chapters 1 and 2). Plants also benefit from the biological environments surrounding them. Microbes, particularly bacteria and fungi form mutually beneficial relationships with plants (see Chapter 2). Increasingly, these relationships are being recognised and translated into forms of biological control against damaging pests and pathogens. This also increases the means by which nutrients are made available for uptake by plants.

Carbon dioxide and water are essential for photosynthesis using energy extracted from sunlight (see Chapter 1). Photosynthesis initially constructs short chains of carbon atoms (see Underpinning knowledge 1.1) which are then further transformed into more complex substances such as sugars, carbohydrates, fatty acids, lipids, amino acids, proteins and nucleic acids. The latter are the building blocks of DNA and RNA, which are the fundamental basis for information exchange between generations and the transfer of genetic instructions resulting in germination, growth, reproduction and senescence (Chapter 3).

The nature and use of light

Natural light provides energy for photosynthesis and reaches earth from the sun (see Chapter 1). Visible light is only part of the energy present in sunlight but it is the segment used by plants, particularly the violet-blue and red portions of the spectrum. Light energy is absorbed by pigment molecules of which two types are involved in photosynthesis in green plants; these are the chlorophylls and carotenoids. Two forms of slightly different chlorophyll are involved in photosynthesis. Most of the energy is collected by "chlorophyll a" from the violet-blue and far-red parts of the spectrum. "Chlorophyll b" and carotenoids utilise energy from the blue, green and red regions of the spectrum. The energy harvested by chlorophyll from light is transferred through a chain of high-energy particles, as described in Chapter 1, which move it into a collection centre in the chloroplast. Photosynthesis is only possible in green plants because of the intricate manner by which energy is gained, transferred and ultimately used in the manufacture of increasingly complex products such as carbohydrates, proteins, lipids and nucleic acids. The enzyme with the shortened name "rubisco" (see Underpinning knowledge 1.1) helps build these increasingly longer chains of carbon and hydrogen atoms, forming sugars which are exported from the chloroplast into the cytoplasm of the cell. There they are used in turn as building blocks in the manufacture of the more complex high-energy molecules.

Using stored energy

Oxygen is required for respiration, which reverses photosynthesis (see Underpinning knowledge 1.2 and Figure 1.9). Respiration breaks down carbon compounds such as carbohydrates and sugars using oxygen as an energy-source releasing carbon dioxide and water (see Chapter 1). The release of energy stored in complex carbon (organic) molecules drives germination, growth, reproduction and senescence. Because of photosynthesis green plants are capable of "self-feeding" (autotrophs) since they utilise and store energy obtained from sunlight. That energy is in turn utilised in respiration. Animals including human beings are "fed by others" (heterotrophs) and require sources of food from which energy can be extracted. Consequently, plants supply and store energy (in physical terms, this is potential energy) which they and the animal kingdom use.

Harvesting the sun

Harvesting the sun is an elegant term for gardening coined by our New Zealand colleagues, Professors Errol Hewett, Ian Warrington and Dr Chris Hale, for the International Society for Horticultural Science (ISHS). Gardening "harvests the sun" using and storing energy for immediate and future use. The chemical bonds between atoms forming molecules such as sugars are sources of stored energy. Energy released by respiration is used in undertaking work. For plants, work is expressed by forming flowers or in confronting pests and pathogens. Each activity requires a supply of energy. Oxygen absorbed from the air reacts with carbohydrates, releasing carbon dioxide, water and energy. This is aerobic respiration because oxygen is involved. Inside cells, energy drives motion such as movements that occur during cell division in mitosis and meiosis and in the passing of genetic information into newly forming cell nuclei. The central importance of phosphate atoms in energy transfer should indicate to gardeners why this element is so important as a constituent of fertilisers.

Waterways

Water is a vital component in all cellular activities because it is the medium in which reactions take place and in which their products are transferred throughout the plant. Soil is the major source of water for plants and is taken in through the root hairs (see Underpinning knowledge 4.1). Carbohydrates for example, are moved in solution in the phloem cells throughout plants into the stem and root apices where cell division and growth require energy for work. Mineral nutrients present in soil are transported upwards through plants dissolved in water. Nutrient transport takes place in the central vascular system, the xylem (see Underpinning knowledge 4.1) within narrow-diameter tubes extending from the roots to the uppermost parts of plants. In the roots, stems, leaves, flowers and fruits of plants, water moves out laterally from the xylem across living cells drawn by their differential concentrations of nutrients.

Water exits from the leaves via stomatal pores. The opening and closing of these pores regulates water loss, controls desiccation and the essential admission of carbon dioxide for photosynthesis and oxygen for respiration. Each day the warming effect of the sun heats the surfaces of leaves leading to the loss of water by transpiration from the stomata. As a result, an increasing suction force develops which is transmitted in the xylem down to the roots resulting in an inflow of water from the soil. At night, when the atmosphere cools down, water loss by transpiration is reduced and inflow from the soil slows or ceases. The stomata are surrounded by guard cells; their function is manipulating the opening and closing of each stoma reacting to the normal daily cycle during daylight and darkness. When the relative humidity around plants is high, there are occasions when the roots continue taking in water and exporting it into the leaves. This leads the leakage of water from glands (hydathodes) at the tips of xylem vessels around the edges of leaves as droplets, a process known as guttation.

Stomata also react when water shortages develop for example, in dry periods or extended droughts. This is the first line of defence against either damaging water shortages developing in the soil or air temperatures rising beyond the normal range to which plants are fitted by evolution. The movements of the guard cells require energy for work in the processes of opening and closing. Potassium is used in developing gradients of concentration between cells, which allows water movement and the transport of other nutrient materials. The guard cells are activated by environmental stimuli such as carbon dioxide concentration, light, temperature and relative humidity.

Plants originating from temperate regions have evolved strategies which combat adverse conditions. Deciduous trees, shrubs and herbaceous perennials shed leaves at the onset of autumnal chilling, resulting in processes of acclimatisation (known as acclimation in North America) which lead on to dormancy during winter. Evergreen plants frequently produce very thick leaves, often with the stomata only sparsely present on the under surface and surrounded by mats of hairs. These minimise water loss particularly in those periods when temperatures are very low and the soil is frozen and water cannot be taken up by the roots. Plants originating from hot, arid regions may have stomata growing in pits in the leaves or stems where the relative humidity is higher than that of the surrounding atmosphere, this helps prevent water loss. Plants originating in the hot, wet tropics develop relatively little defence against water loss, and consequently when they are used in the garden are very susceptible to damage during periods of water stress. By contrast, plants originating in hot, dry, tropical zones have evolved substantial mechanisms for defending against water deficits such as are found with the cacti. These plants have sparse numbers of stomata, and frequently these are deeply seated in ridges which retain humidity, preventing water loss. Some desert cacti also have large spines and hairs as defence mechanisms against browsing herbivores and as means for coping with the large diurnal variations in temperature (see Figure 4.60).

Figure 4.60 *A cactus showing hairs and spines, providing protection from herbivores and water conservation*

Most nutrients especially those required in large amounts are actively taken in through the root cell walls. This process requires energy which comes from respiration in the root cells using oxygen taken in from the soil. This is one reason why gardeners should maintain well-aerated soil structures. Soil which is waterlogged or compacted and degraded by trafficking will have only limited supplies of oxygen, preventing healthy root growth. This restricts the uptake of nutrients, resulting in the development of starvation or deficiency symptoms in the stems, leaves and flowers.

In healthy plants, mineral nutrients move into the xylem stream and are relocated by the transpiration stream. Macro-nutrients such as nitrogen, phosphorus, potassium and calcium are required in relatively large amounts for growth and reproduction (see Underpinning knowledge 2.2). Other micro-nutrients have specialised functions. For example, magnesium is a central component of chlorophyll, sulphur is a constituent of specialised amino acids, boron aids enzymic reactions and is involved in the linkages that bond the cell wall structures and in the formation of some growth-regulating hormones.

Sources and sinks

The phloem distributes manufactured products from photosynthesis by translocation. During active growth phases some products may be excess to immediate needs and are diverted into storage. Storage structures range from either fleshy or dried fruit to specialised swollen organs such as the rhizomes of *Iris* or *Dahlia* tubers (see Chapter 7). Carbohydrates are translocated from the point of production, the source, and placed where they will be used or stored, the sink. During translocation carbohydrates are in soluble forms such as various sugars. These are loaded into the phloem by an active transport process for which ATP provides the energy which drives differentials in concentration between receiving and delivering cells (see Figure 4.61).

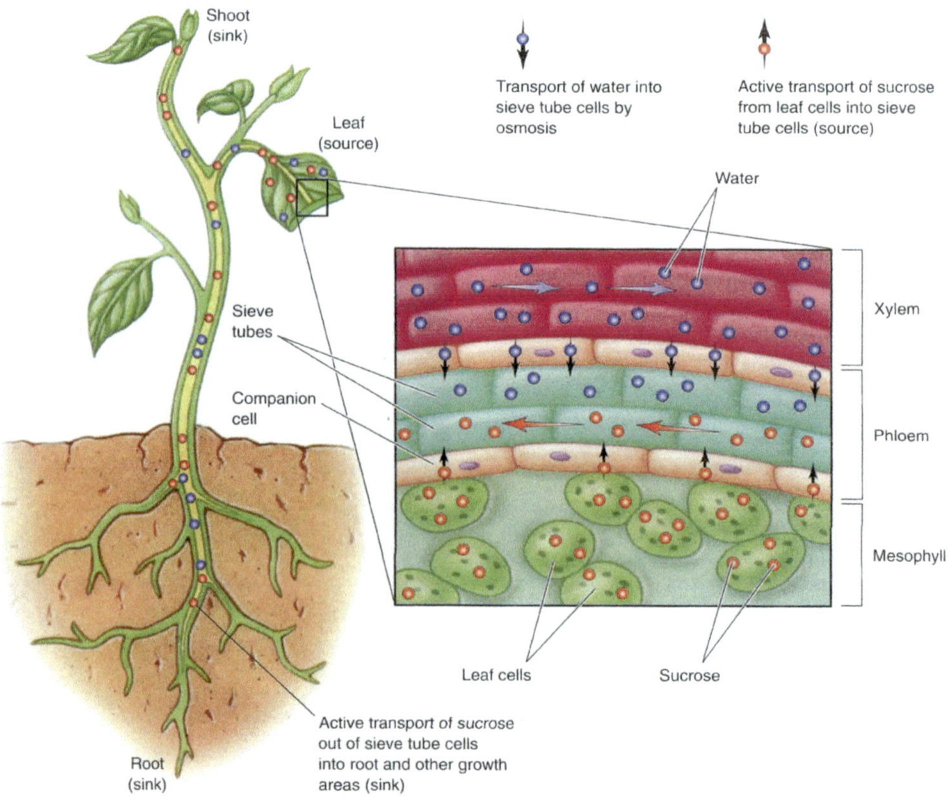

Figure 4.61 *Sources and sinks*

Removing resources at the point of use is again an active unloading process using energy which drives the work.

Helpful microbes

Soil is a living entity which contains huge numbers of helpful and beneficial microbes in addition to its other attributes. Increasingly, it is appreciated that biological components of the aerial and soil environments support active, healthy growth by green plants. Microbes such as bacteria and fungi live in association with roots and leaves (see Underpinning knowledge 2.3). Soil health is judged by the level of activity from helpful or benign microbes. Helpful microbes in soil are generally referred to as "benign" whereas unhelpful forms, which cause disease, are referred to as "pathogens". The rhizosphere zone, around roots, contains enormous numbers of beneficial microbes. These benefit from sugars and some growth-promoting substances exuded from roots possibly because they are excess to requirements. The microbes help by defending the roots from damage caused by pathogenic microbes or predatory insects. Natural biological control mechanisms are gradually being understood in greater depth and with increasing appreciation of the way healthy growth and reproduction of green plants are boosted. Soil health is encouraged by providing ample aeration and drainage which oxygenates the roots. Adding farmyard manure or well-rotted composts, containing beneficial microbes, promotes soil health and continual applications over several years increases earthworm activity. Worms are "Nature's plough", as Charles Darwin asserted, and their burrowing, digestion and excretion of soil particles increases aeration and drainage. Soil health is increased further by growing legume crops like peas and beans (Chapter 3). These have natural associations with beneficial bacteria which increase soil nitrogen reserves by capturing nitrogen gas from the atmosphere and converting it to soluble nitrates and nitrites which they and plants can utilise.

Crops such as some of the alliums, garlic, leeks or onions, have large vigorous root systems which physically improve soil structure. Additionally, they release compounds containing sulphur from their roots which are antagonistic towards some parasitic and pathogenic organisms; this process is now being exploited by gardeners and is known as bio-fumigation. Most plant root systems, apart from members of the *Brassica* family form direct relationships with beneficial fungi. These develop either as sheaths wrapped round individual roots or as invasions and colonisation of the outer layers of cells. These relationships are termed as ecto- or endo-mycorrhizae respectively (see Underpinning knowledge 2.3). Although known for over a century, only now, with the aid of molecular-biological techniques, is the depth of these valuable and subtle relationships being appreciated and commercial preparations of them being made widely available. Mycorrhizae provide biological control of pathogens and sources of otherwise inaccessible phosphorus and water. Although most soils are apparently well supplied with phosphorus, it may not be easily accessible for plant roots, especially during periods of rapid growth such as when potatoes are forming new tubers and these are swelling.

4.3 Weeds, pests and pathogens

Cultivating plants is a process in which ideal growing conditions are created. Inevitably, this attracts competitive organisms for which gardens are ideal places for their growth and reproduction.

A dispiriting aspect of gardening is when a lush and vigorous vegetable or fruit crop or very floriferous flower display is just reaching peak perfection and suddenly it is attacked by pests, pathogens or smothered by weeds. Experience, anticipation and planning should make most such events avoidable. But there will always remain times when "fire-brigade" action is needed because an alien insect or pathogen has entered the garden unexpectedly. As one very notable Scottish weed scientist, Harry Lawson, observed "weed control is a moving target", and this applies equally for pests and pathogens.

Weeds

Serious weed competition usually results from inadequate soil preparation ahead of sowing or transplanting. Weeds are of two general types, perennials or annuals. Perennial weeds such as couch grass (*Elytrigia repens* also referred to as *Agropyron repens*), ground elder (*Aegopodium podagraria*) (see Figure 4.62), docks (*Rumex* spp.), nettles (*Urtica dioica*), and thistle (*Cirsium arvense*) must be removed well before cropping begins.

The most effective means of removal of these weeds is by careful, deep digging and removing every last fragment of weed roots. The resultant green waste is best sent to the nearest municipal dump for composting. Garden compost rarely reaches temperatures which will destroy effectively perennial weeds. Normally removing perennial weeds is part of the annual cultivation cycle starting in the late autumn and continuing into winter. Where the garden is part of a new build or has been neglected for some years, then deep digging and weed removal are essential and should be started at all times of year (Chapter 2).

Active perennial weed growth is best controlled by spraying with total herbicides such as glyphosate. This herbicide is the subject of much debate as to its use; currently, it continues to be accepted for use in Europe provided the instructions on the label are followed. Weed growth can be weakened in advance of spraying by covering the garden area with black polythene sheeting. The weeds grow under the plastic but because light is excluded the plants become yellowed and stretched (etiolated) (see Figure 7.35). Additionally, the polythene traps heat and warms the soil encouraging this weakened growth still further. Spraying these stressed shoots reduces their vigour even more. Nonetheless, meticulous digging over with a garden fork and sifting out every last piece of root is still essential. Otherwise, residual root fragments will grow and thrive without competition. Growing a useful and vigorous crop such as potatoes on the entire area in the first season will help by smothering resurgent weeds once the foliage canopy has closed over. When the ground is dug again after cropping, it is vital that residual weed roots are picked out and disposed of away from the garden.

Figure 4.62 *Ground elder (*Aegopodium podagraria*), showing that the subterranean rhizomes provide a capacity for extensive colonisation*

Annual weeds are those which germinate and grow in one season producing crops of seeds and then die-down. Some annual weeds are capable of rapidly reproducing several generations of seed during one growing season and are known as ephemerals, for example chickweed (*Stellaria media*), scentless mayweed (*Tripleurosperum inodorum*), groundsel (*Senecio vulgaris*) (see Figure 4.63) or fathen (*Chenopodium album*) (see Figure 4.64).

Weed competition for light, nutrients, water and space presents the gardener with considerable problems especially when crops are in the seedling stages. Competition reduces the vigour and cropping potential of all vegetables, damages the productivity of soft fruit and the enjoyment of ornamentals. There are "critical periods" when weeds dominate and damage the growth of crop seedlings more severely and that increases damage. When crops produce full leaf canopy closure, weed growth is suppressed. But in crops like onions, leeks and garlic (alliums) where the leaves grow vertically without closing the canopy, weed damage continues until harvesting. Particularly in allium crops weed competition is linked with the formation of "thick necks" as the stressed plants produce flowering heads and the resultant bulbs cannot be stored. Weeds are havens for pests and pathogens which then infect crop plants and that further increases resultant damage.

Removal of the seedling weeds is essential and achieved by hand cultivation. Weed control is aided by using a good-quality, very small, sharp hoe which cultivates the land between crop rows cutting-off weed seedlings from their roots and leaves them drying out. The aim should be total weed removal, even one or two residual plants will flower and seed storing-up trouble for future years. The gardening adage "one year's seeding is seven years weeding" is very accurate. Some vegetables can be sown or transplanted through plastic sheeting, which further helps suppress weeds.

Similar principles of weed control apply for soft fruit. Strawberry, raspberry and blackcurrant beds should be kept free from weeds (Chapter 5). Weed competition quickly reduces the

Figure 4.63 *Groundsel (*Senecio vulgaris*), an ephemeral annual weed capable of producing several generations in one season*

Figure 4.64 *Fathen (*Chenopodium album*) produces large numbers of seed clusters on each plant which resemble the seed of beetroot*

productivity of strawberries. Infestation, particularly of perennial weeds into longer-term crops like blackcurrants and raspberries, has similarly deleterious effects. Removing perennial weeds becomes more difficult once the bushes are mature.

Pests and pathogens

Most garden pests are insects and infections are predictable. Aphids (black bean aphid, *Aphis fabae*, or peach-potato aphid, *Myzus persicae*), caterpillars (large cabbage white butterfly, *Pieris brassicae*, or small cabbage white butterfly, *Pieris rapae*) (see Figure 4.65), cabbage root fly, *Delia radicum*) and carrot fly (*Psila rosae*) are regular enemies.

Most pathogens are fungal or allied organisms, but because of climate change, bacteria which require warmer and wetter conditions are becoming more frequent causes of disease.

There are several means of control:

Natural bio-control

Gardens contain mixtures of plant types and it is probable that natural enemies of pests are present and should be encouraged. Ladybirds (various genera and species of the family *Coccinellidae*, coccinellids) (see Figure 4.66), for example, offer bio-control of aphids.

Mixing beneficial flowers such as marigolds (*Calendula*) which encourage populations of ladybirds and other predators into the vegetable garden is a helpful strategy. These plants in particular exude chemicals from the roots which reduce the activities of pests and some pathogens. This is known as "companion planting". Forms of commercial bio-control which control insects and some molluscs such as slugs and snails may be purchased from garden centres and by mail order. Beneficial fertiliser products such as calcium cyanamide provide sources of nitrogen and soluble calcium, and in addition encourage benign microbe populations in the soil which attack and degrade soil-borne pathogens. Calcium cyanamide and the more soluble calcium nitrate have

Figure 4.65 *Caterpillars of cabbage white butterflies (*Pieris *spp.)*

Figure 4.66 *Ladybird (*Coccinella septumpunctata*)*

very valuable properties for vegetable and fruit gardens beyond simply providing sources of nutrients.

Crop rotation

Rotation is a first step in ensuring that soil-inhabiting pests such as cabbage (*Delia radicum*) and carrot root flies (*Psila rosae*) or pathogens such as clubroot (*Plasmodiophora brassicae*) (see Figure 2.54) or onion neck rot (*Botrytis* spp.) (see Figure 2.55) and white rot (*Sclerotinia* spp.) are kept in check. In crop rotation, groups of vegetables such as potatoes, legumes, brassicas and small-seeded crops follow each other in sequence over a four-year period. This is an important part of careful husbandry (see Table 2.1). Regular liming of the land is useful for diminishing the effects of clubroot (*P. brassicae*). Care is required because there is a pathogen that attacks potatoes causing common scab (*Streptomyces scabies*), for which liming soil may encourage infection. Consequently, lime should be applied after potatoes in the rotation, and as a result soil pH will have become more acidic when they return in the rotational sequence (see Underpinning knowledge 2.4).

Resistant varieties

Seed of resistant varieties of carrot and cabbage is very effective in controlling some insects and soil-borne pathogens. Genetic resistance is the simplest, most effective and environmentally benign form of defence but rarely offers complete control and when used for several seasons gardeners may find that efficacy is diminishing. This is because parts of the pest or pathogen population which have evolved capabilities for invading the resistant variety have become dominant, this is known as the "boom and bust cycle".

Environmental control

Vegetable seed should be purchased from a reputable seed house. Home-saved seed is a false economy since it may be infected with a galaxy of pathogens and possibly some pests. Seedling-invading pathogens become apparent in the propagation stages, particularly damping-off diseases (see Figure 2.41). Several fungi invade the stem bases of seedlings, causing watery rot and death. Generally, these problems are avoided by careful watering and providing ample ventilation in the greenhouse or cold frame. It is essential that composts used for seedling propagation drain freely and provide effective root aeration.

Avoidance

Cleanliness and tidiness are important means for avoiding pests and pathogens. Removing weeds and senescent foliage, collecting fallen leaves, washing and oiling tools helps prevent invasion and infection. Skilful gardening husbandry is also important. Broad beans, especially the overwintered crops, for example, are very attractive for aphids, which can detect the tops of these crops by their green hues which differ from those of the main canopy (see Chapter 3). Consequently, removing the top shoots from the plants as pod swelling begins reduces the risk of invasion (see Chapter 3, and Figure 3.38). Aphids cause direct damage and they are carriers of viruses which are injected into the plant when the insects feed; excluding these carriers provides several benefits. Placing collars or a ring of coarse sand around the necks of cabbage and other brassica plants helps deter adult cabbage root flies (*D. radicum*) from placing their eggs close to crop plants. Very fine mesh netting placed over brassicas deters adult cabbage white butterflies (*P. brassicae*, and *P. rapae*) from egg-laying on the under-surfaces of leaves, reducing crop defoliation, and keeps the maturing heads free from deeply seated caterpillars and their frass (an alternative term for faeces) deposits. Netting also deters birds such as pigeons, while traps may be required as a means of eliminating mice which find the sugars in maturing carrots and beetroot, for example, very attractive.

Using the correctly designed composts for propagation will also aid the avoidance of diseases. Composts used for germination must be formulated such that there is adequate drainage. Potting plants with well-developed roots requires composts containing an adequate supply of nutrients so that further growth may continue. The absence of sufficient nutrients halts growth and leads to deficiency symptoms.

Sprays

The range of insecticides and fungicides available for gardeners has been reduced by more than 50 per cent in the last decade as a result of increasing legislative controls. Those still offered have undergone intensive testing and assessments determining their safety for humans, wildlife and the environment. Preparations available for gardeners are considerably more dilute than those used commercially. When selecting preparations for use, read the labels carefully and adhere to all the precautions specified by the manufacturer. When used safely, sprays offer a quick and very effective means of dealing with difficult pests and pathogens. Potato blight (*Phytophthora infestans*) spores, for example, can arrive just as crops are maturing and completely ruin them (see Figure 2.33). Monitoring plants at regular intervals and applying treatments at the first signs of invasion is essential.

Soft-fruit plants should always be purchased from reliable garden centres or mail-order specialists and be accompanied by health certificates. This avoids many of the pests and pathogens which attack these crops. Nonetheless, preventing the disfigurement of strawberries by aphid (*Chaetosiphon fragaefolii*), raspberries by cane midge (*Resseliella theobaldii*) whose larvae eat the developing canes internally, and raspberry beetle (*Byturus tomentosus*) where the larvae colonised the inside of fruits, requires regular crop monitoring and spraying at the first signs of invasion. Blackcurrants are affected by Big Bud mite (*Ceridophyopsis ribis*) that bores into the vegetative buds causing swelling and deformation. The only solution is pruning out affected shoots and disposing of them away from the garden.

4.4 Food shortages and excesses

Shortages (deficiencies) of nutrients and occasionally excesses (toxicities) are usually the result of inadequate, or over-generous, use of fertilisers. Plants require a balanced and continuing supply of macro- and micro-nutrients from the soil or other substrates (see Chapter 2). Nutrient shortages, or less frequently excesses result in plants becoming unthrifty and showing yellowing, browning, blackening or other symptoms in the leaves, stems, flowers and fruit.

Symptoms of nutrient deficiency or toxicity can resemble those caused by pests and pathogens. But they lack signs of predation such as colonies of aphids or caterpillars and feeding damage or the presence of fungal sporulation and rotting. The absence of these signs should provide evidence that discolouration and deformity are being caused by a lack of a particular nutrient.

Applying farmyard manure or well-rotted compost when the land is dug is the first line of defence against nutritional failures. The second line is using lime so that the soil pH is maintained at about neutral or slightly above (pH = 7.0 to 7.4), as discussed in Chapter 3. Raising the garden pH above these values can also cause nutrient deficiencies, particularly "lime-induced chlorosis" where excess calcium inhibits the uptake of magnesium or iron. This is because calcium preferentially enters the roots excluding the other elements. Adding granular fertilisers at the time of sowing or planting is the third line of defence. Fertilisers are usually quick-acting because they contain highly soluble ingredients, making the nutrients readily available for the root systems. Fertilisers can also be applied into growing crops as top dressings when the end of first phase of rapid growth has been reached. This avoids adding all the fertiliser at sowing or planting when a proportion cannot be immediately used by the crop and will probably be washed by rainfall away from the root zone into drainage water and wasted. Top dressing adds specific highly soluble nutrients such as nitrogen and potassium into crops boosting ultimate yields. Nutrients may also be added during crop growth in liquid form. Adding nutrients in this way increases the speed of uptake because they are readily available in solution for the roots. That is particularly important during periods of water shortage.

Recognising signs of nutrient shortage

Recognising the precise cause of nutrient shortages requires, knowledge, experience and probably the services of a soil and plant analysis laboratory. Nonetheless, some basic guidance particularly for the major nutrients may be helpful. Nitrogen deficiency is the most common cause of problems in vegetables largely because of their speed of growth which can exhaust supplies in the soil. At the seedling stage, the growth of nitrogen-deficient plants is slow and the leaves are smaller than would normally be expected (see Figure 2.49). Nitrogen deficiency in adult plants causes development of a pale green colour in the leaves, which may drop off sooner than expected.

As deficiencies increase in severity, leaves turn vividly purple, red and orange in colour particularly with brassica crops. Applications of quickly absorbed forms of nitrogen such as potassium or calcium nitrate may help correct this deficiency. Seedlings suffering from nitrogen deficiency may be watered with a very dilute solution of this fertiliser. This should be done cautiously, preferably by testing the concentration first on a sample of plants. Concentrations which are too strong will scorch or even kill seedlings. These fertilisers can be obtained from a garden centre or mail-order supplier.

Both potassium and phosphorus deficiencies cause symptoms which at least initially resemble shortages of nitrogen. Affected seedlings show slow and poor growth with a lack of green lustre in the leaves. Phosphorus deficiency can develop in very rapidly growing crops.

Some forms of soluble phosphate, known as Struvite, are entering the commercial market but have yet to reach garden suppliers. Deficiencies of potassium may happen as this fertiliser is soluble and easily washed through the soil. Symptoms include the development of rosette forms of growth and an inability for stem extension. Potassium deficiency is not uncommon in potatoes causing speckling on the leaves. This is because the large amounts of root and tuber growth produced by potato crops for which this element is essential and the deficiency symptoms are then expressed in the leaves. Solutions of potassium nitrate should be watered around affected plants. This fertiliser may be obtained from garden centres and mail-order suppliers.

Accurately diagnosing the precise cause of other nutrient deficiencies usually requires specialist advice. Some crops such as lettuce develop scorching of the leaf tips which may result from shortages of calcium. In this case, quick-acting remedies include applying calcium nitrate or calcium chloride solutions watered round the plants. Regular liming after digging is a more successful solution but results of applications of lime as calcium carbonate only become effective after several months. Slaked lime is a form in which calcium carbonate has been heated and then treated with water forming a very fine powder. It is quicker acting that calcium carbonate but has the disadvantage of being caustic and hence should be applied with considerable caution (see Chapter 2).

Magnesium deficiency damages the structure of chlorophyll molecules and causes yellowing between the veins of leaves particularly the main and secondary ones. Extreme deficiency causes total yellowing and browning of leaves which fall-away. Applications of magnesium sulphate (Epsom salts) are a short-term treatment for established crops. A more durable one is incorporating lime derived from magnesian limestone rather than calcium limestone during autumn and winter digging. Iron deficiency can be confused with symptoms caused by shortages of magnesium, but the uniform yellowing of younger foliage is a particular sign of iron deficiency. Iron deficiency is associated with calcareous soils with alkaline soil pH values. This deficiency is worst with cold, wet, heavy clay soils where root growth is slow. The use of chelated fertilisers as top dressings or foliage sprays and applications of iron sulphate fertiliser will help alleviate the problem (see Underpinning knowledge 2.4). Chelates are a chemical formulation in which atoms of metals such as iron are held in a claw shape structure (hence the modern Latin name *chela* which itself is derived from the Greek *chele*) from which they are released, helping alleviate calcium-induced deficiencies. Deficiencies of other trace elements such as boron, copper, manganese and molybdenum are found infrequently. The classical symptom of molybdenum deficiency is whiptail in cauliflower and broccoli. The leaf lamina becomes twisted and much reduced, leaving only the midrib. Excessive liming of soils beyond pH 7.5 may induce molybdenum deficiency. Treating the soil and spraying plants with sodium or ammonium molybdate helps alleviate this deficiency and prevent its recurrence.

Excessive concentrations of nutrients are most frequently caused by over-generous applications of fertilisers before sowing or as top dressings. This results in excessively dark green leaf growth especially in brassicas. Affected plants are unstable and over-large with weak stems and frequently topple over. Leaves containing excessive nitrogen are more disease prone and cabbage heads, cauliflower curds and Brussels sprout buds take on a bitter flavour when eaten.

Treatments include avoiding excessive nitrogen applications especially fertilisers containing ammonium molecules. Excessive fertiliser applications may result in white salty deposits appearing on the soil surface (see Figure 2.52). Affected soils should be cultivated and then left until significant rainfall has washed the excess fertiliser away.

Toxicities from heavy metals such as boron, copper, manganese result from industrial pollution which is a declining event in Europe. Where these toxicities develop they cause excessive yellowing and possibly scorching of leaves. Flooding land may result in chlorine toxicity resulting from large quantities of dissolved salts present in the water. There are no feasible remedies for these toxicities except allowing natural drying and draining of the soil. Subsequent cultivation requires care since flooding destroys soil structure and, consequently, aeration and drainage are only slowly restored. Rainfall will wash toxic pollutants from the soil and gradually microbial colonisation restores its structure and fertility. Flooded land should be sown with perennial ryegrass (*Lolium perenne*) which, by virtue of its vigorous fibrous root system, encourages the re-establishment of soil health.

Achievements: have you understood and learnt?

At the end of the chapter, you should be able to:

1. Understand the correct storage conditions for packets of small seeds such as beetroot, broccoli (calabrese), Brussels sprouts, cabbage, cauliflower, carrot, leeks, lettuce, parsnip and Oriental vegetables such as Chinese cabbage.
2. Know that small-seeded crops may be sown directly into the garden and alternatively crops like brassicas and lettuce can be sown in seed trays and propagated as transplants.
3. Know the correct preparations for sowing small-seeded vegetable crops and how they should be cultured in the greenhouse or alternative protection, regularly inspected and watered encouraging robust growth.
4. Know that modular plants grown by professional nurserymen are available from garden centres and through mail-order suppliers.
5. Know that frost-susceptible vegetables such as courgettes and sweet corn must be sown in pots filled with compost, transferred to a greenhouse bench or alternative protection and the husbandry required.
6. Know how to transplant courgettes and sweet corn into the garden and apply measures for their protection against slug and insect damage.
7. Know that leeks can provide a valuable garden crop for the late autumn and early winter and that seedlings may be grown in modular trays until they are ready for transplanting or be obtained from a garden centre or mail-order suppliers.
8. Understand that crop plants differ in their form (morphology) and this determines how they will be spaced in the garden.
9. Describe the constituent parts of plants (roots, leaves, stems and buds) and the similarities in cellular construction and functioning describing how the transport of water, nutrients and the products of photosynthesis and respiration is achieved.

10. Understand how differences in the development of crop canopies can be used by the gardener for the purposes of weed control.
11. Understand the role of water within plants providing means for transportation as a result of the transpiration stream operating between the roots and the leaves.
12. Understand the importance of healthy soil, inhabited with benign microbes and animals such as worms leading to the effective functioning of plant roots and subsequent distribution of water, nutrients and the products of chemical reaction within plants.
13. Understand the relationship between soil characteristics and the availability of nutrients for uptake by roots.
14. Understand the impact of weed competition in reducing the growth, quality and yield of vegetables and the differences between perennial and annual weeds.
15. Know the effects of pest infestation on garden plants and how the different types of pest damage can be identified and controlled.
16. Understand the different types of symptoms caused by plant pathogens and their effects on garden plants and means by which plant pathogens may be controlled.
17. Know the importance of nutrients for the growth, quality and yield of garden crops and plants.
18. Understand the varying forms of symptoms which result from deficiencies and excesses of nutrients.
19. Understand the causes of toxicity symptoms resulting from excesses of particular nutrients affecting garden plants.

Definition of terms: *Describe* is able to show why a scientific principle is important; *Understand* applies scientific principles into the context of gardening; *Know* carries through scientific principles in practical gardening activities.

Chapter 5
Growing soft fruit

Introduction

This chapter introduces vegetatively propagated perennial fruit crops which mature over several years from planting to full cropping. This contrasts with earlier chapters which discuss growing annual crops from seed and from vegetative organs such as potato tubers and onion sets. Soft fruits produce edible, juicy, succulent products. In nature, these evolved as colourful, sweet berries which attract birds and other animals who by digesting them, excrete and distribute their seeds. The cultivation of soft fruit still results in products attractive for birds and this causes one of gardeners' problems.

Garden practices

a. Forms of soft fruit

Herbaceous perennials, such as strawberries are characterised by the aerial plant parts growing through each season and then shedding their leaves and entering a dormant or quiescent state in the winter. The roots and rootstock remain viable during winter months and regrow once soil temperatures reach about 5 °C. This is a normal life cycle for herbaceous plants originating from geographical areas with cold winters (see Chapter 7). Dormancy is the mechanism for damage avoidance. The term "rootstock" is that part of a perennial plant consisting of the root system and possibly the lower stem that enter the soil and are united possibly below a graft union in top fruit trees.

Perennial shrubby bushes, such as black, red and white currants, are characterised again by seasonality and plant structures which are protective mechanisms for avoiding the consequences of harsh winter weather. The plants grow forming leaves, shoots and flowers from early spring onwards and insects fertilise the flowers which mature into fruit crops. After fruiting the leaves senesce, fall away and the bushes become dormant. Buds which will produce fruit in the following season develop during flowering for the current year.

Cane perennials include raspberries, blackberries and hybrid berries such as tayberries and loganberries. The canes are long, flexible stems which are strengthened by thickening of cell walls such that they withstand damage in windy environments. These plants evolved using other plants as means of support. In the garden, support is required for the canes because fertile soils encourage considerable extension growth. Traditionally, raspberries form fruit on canes developed in the previous season. These young canes produce considerable foliage during that season which dies in the autumn and dormancy ensues. They reform foliage in the spring accompanied by flowering and fruiting. Coincidentally, new shoots develop which will crop in the following year. Once the two-year-old canes have fruited, they are then removed, allowing the developing new flushes of growth ample space.

In both strawberries and raspberries, plant breeders have developed variations on these biennial themes which permit extensions of the fruiting seasons.

b. Certified plants

Purchasing high-quality perennial soft fruit plants from an accredited source is essential. This is because vegetatively propagated plants accumulate and retain virus and fungal diseases and possibly some insect pests which rapidly reduce their vigour and productivity. Considerable research has been invested in understanding the biology and spread of these diseases and pests. As a result, producing and marketing healthy certified plants are regulated by statutory provisions because of the implications for the commercial fruit industries worldwide. All soft-fruit plants offered for sale must carry evidence of their source, official inspections and declarations of disease freedom. Plants bought from garden centres and mail-order companies should provide documentary evidence that plants are healthy and fit for use. Officially "Certified Plants" will either carry a "Plant Passport" or labels with the number, the source of production and of having been propagated at a *bona fide* commercial source.

c. Growing soft fruit

Each of the perennial soft-fruit types discussed here has a range of varieties, each with differing characteristics. Deciding which ones suit the requirements of a particular garden is an enjoyable winter pastime. Lists are provided by seedsman and other mail-order suppliers either in hard copy or via the web. Useful information is found in the gardening sections of the newspapers, gardening magazines and *The Garden* published by the Royal Horticultural Society of London (RHS). Most beneficial are conversations with neighbours and fellow gardeners in local gardening clubs and garden centres. Local knowledge is very valuable because variety suitability varies substantially within even small geographical distances.

Frost and low temperature tolerance is important for all types but especially blackcurrants. Older varieties such as 'Wellington XXX' and 'Baldwin' (the age of these varieties can be judged from their names) are frost-susceptible, low temperatures kill the flowers coinciding with periods of bee pollination which then fails, resulting in poor fruit set and low yields. Varieties in the 'Ben' group bred at the James Hutton Institute, Dundee, have delaying flowering, which alleviates this problem, making fruiting more reliable, particularly 'Ben Tirran' and 'Ben Connan'. The variety 'Ben Hope' is reputedly resistant to Big Bud mite (*Ceridophyopsis ribis*). Breeders have increased the berry size and sweetness of 'Big Ben' and 'Ebony'. Gardeners will also find some new varieties coming onto the British market from Scandinavia and Eastern Europe, such as 'Cassisima', 'Black Marble' and 'Ebony'. Blackcurrants have been very popular fruit in these countries for many years.

Strawberry production commercially is a huge worldwide industry aiming at supplying fruit into supermarkets year-round. Consequently, there is a very large range of strawberry varieties, and some have been raised in such a way that they will fruit outside the normal summer season. Specially treated runners with accelerated fruiting are offered for gardeners by mail-order companies and some garden centres. Individual preference is the deciding factor here and gardeners must consider whether or not they wish for fruiting outside the normal summer season. Treated plants may revert to the normal growth and fruiting cycles after the initial season although some varieties are genetically programmed for earlier or later fruiting. When purchasing strawberries, the gardener should check carefully the varietal characteristics.

More traditionally, the cropping season may be extended by growing varieties which vary in their season of maturity. Early season (June to early July) varieties include: 'Christine', 'Honeoye', 'Korona' and 'Vibrant', mid-summer types (mid-June to mid-July) are 'Elegance', 'Elsanta', 'Hapil', 'Malling Centenary' and 'Sonata', with late season (mid-July to August) including: 'Fenella', 'Florence', 'Malwina' and 'Symphony'. Some varieties are termed "perpetual or everbearing", fruiting from mid-July to October. Gardeners may wish to use "heritage" varieties such as 'Royal Sovereign', 'Cambridge Favourite' and 'Red Gauntlet' which have a long history of production but may suffer from problems with diseases, especially grey mould (*Botrytis cinerea*) infection (see Figure 5.15). Other varieties, such as 'Sweet Colossus' and 'Vibrant' are available depending on where gardeners purchase their plants. Also available are some highly flavoured "French strawberries" sold as cold-stored runners such as 'Gariguette', 'Mara des Bois' and their progeny 'Manille'.

Raspberries normally require a full growing season followed by winter after planting before fruiting starts, these are termed floricane types. Alternatively, there are primocane types which fruit on the current season's growth in the autumn (late July to October), these include red-fruited varieties such as 'Autumn Bliss', 'Paris', 'Polka', 'Joan J' and the golden-coloured fruit of 'All Gold'. Some floricane varieties are offered as "long extra vigorous canes" of about 1.5 m in length. These canes have selected by the commercial propagators as possessing extra vigour and with the potential for fruiting, albeit modestly, in the year of planting. Thereafter, they revert to normal growth cycles. Normal canes will have been pruned by the commercial propagator to about 10 cm in length. Varieties of floricane raspberries include 'Malling Juno' (June to July), 'Glen Ample' (late June to mid-July), 'Glen Clova' (July), 'Tulameen' (early July to early August). 'Glen Coe' which is a floricane type produces fruit which are deep purple in colour with intense flavour borne on clumps of canes, a plant habit well suited for the smaller garden. The variety 'Ruby Beauty' is suitable for planting in pots and other containers for patio production. Reputedly, fruit that are purple or golden in colour are more beneficial because they contain higher concentrations of antioxidants. Some ever-bearing types with long fruiting seasons such as 'Versailles' are available from mail-order suppliers.

d. Purchasing and pre-planting care

Over-enthusiastic purchasing should be avoided; 10 to 12 strawberry plants, 1 or 2 blackcurrant bushes and 6 to 10 raspberry canes will suffice for most small gardens. These will provide sufficient strawberries and raspberries for two people. Blackcurrant bushes are very vigorous, and after 3 to 5 years a couple of plants will yield ample fruit for the table plus small amounts for jam-making.

Unwrap the plants immediately on arrival removing all rubber bands and packaging. Tease out the root systems and place them in a bucket of water and stand this in a frost-free garage, shed or greenhouse. Immerse the plants so that all the roots are covered up to the point where leaves emerge from the rootstock. It is of paramount importance that the root systems are kept damp until they are planted. Root desiccation is one of the main reasons for the failure of soft fruit plants. If planting is delayed for more than a couple of days, the plants should be transferred from water into a bucket containing moist compost. The Scottish raspberry growers' adage that "dry roots are dying roots" is very true.

e. Strawberry (herbaceous perennial soft fruit)

Strawberry fruits are one of the most popular garden delicacies (see Figure 5.1).

The British growing season can now be extended from March through to October with crops grown under polythene tunnels as well as in the open. But true strawberry freshness, flavour and lusciousness comes from garden-grown and picked fruit which is eaten almost immediately after they have been picked. For this reason, gardeners should find growing small quantities as an early summer fruit attractive.

As preparation for planting, separate the strawberry plants from one another. Identify the crown, or growing tip, of each plant where residual leaves and some buds meet with the root system (see Figure 5.2).

Figure 5.1 *A basket of succulent strawberry fruit*

Figure 5.2 *Strawberry bare root transplants*

Inspect each plant, ensuring that it is free from obvious pests or diseases. Wrap the roots in wet newspaper or hessian sacking while they are waiting for planting (see Figure 5.3).

Level out and cultivate an area of previously well-dug and manured land incorporating "GrowMore" fertiliser at 30 g per square metre using the garden rake. Cultivation should aim at producing a tilth which contains small clods of earth up to about 3 cm in diameter. For strawberry planting, a very fine seed bed tilth is unnecessary. Make a straight line across the area using a garden line as a guide for planting (see Figure 5.4).

For each plant, make a hole with a trowel that is sufficiently large that the entire strawberry root system may be buried to a depth of 3 to 4 cm and mix in some multipurpose compost (see Figure 5.5) and moisten well (see Figure 5.6).

The roots of each strawberry plant should be carefully arranged in the hole, ensuring that they do not crowd each other or overlap. That ensures they have spaces for growth and can penetrate more easily into the surrounding soil (see Figure 5.7).

Figure 5.3 *Strawberry transplants being kept moist to avoid drying out*

Figure 5.4 *Preparing a planting hole with a trowel*

Figure 5.5 *Mixing compost into the planting hole*

Figure 5.6 *Moistening the soil and compost in the planting hole*

Figure 5.7 *Placing the roots of the transplant in a hole*

Figure 5.8 *Covering the roots with soil*

Figure 5.9 *Water the transplant well*

Figure 5.10 *A well-established strawberry transplant*

Place each plant in its own hole and move soil back around the neck of each plant. Raise the plant a couple of times so that soil is in close touch with the roots and then gently firm it with your fingers. The neck of the plant, where the leaves exit from the stem base should be just visible above soil level (see Figure 5.8) and water well (see Figure 5.9). This equates with the position of the plant in the original propagation bed.

Each strawberry plant requires growing space separating it from its neighbours by about 30 cm in each direction. This provides room for growth and permits husbandry operations such as adding nutrients and covering the plants when fruiting (see Figure 5.10).

Sprinkle slug bait round the strawberries after planting and add more at intervals. Alternatively, before planting treat the land with a bio-controlling nematode preparation which may be obtained from a garden centre or by mail order. Using biological controls are considered as helping retain song-bird and hedgehog populations in the garden. Ripening strawberries are an attractive food for slugs and snails.

Laying black polythene on the soil surface once it has been cultivated and before planting the strawberries will help conserve soil moisture around the strawberry roots and prevents weed growth competing with the crop. Bury the edges of the plastic in the soil by taking out a "V"-shaped trench into which edges of the sheet are placed and bury these with soil. The sheet should be as tight as possible preventing winds from causing damage. The strawberry plants are inserted through the plastic in a similar manner to that described above (see Figure 5.11).

Some strawberry plants will fruit in their first year after planting because they have been given a period of cool conditions before they are sold, these are called "cold-stored runners". The gardener should identify the potential fruiting season of plants from the supplier before purchase.

Some plants which are bought and established in the early spring may also show some precocity of fruiting with a sparse crop of berries in the first summer. These flowers should be removed as soon as buds emerge allowing the plants opportunities for building strong and vigorous growth for future cropping years, although one or two might be left allowing the gardener opportunities for testing the variety's flavour. Flower stalks are easily removed by nipping them off manually between the thumbnail and first finger or snipped off with sharp secateurs, removed completely and disposed of.

Normally, strawberries fruit in the season following planting (see Figure 5.12).

Fruiting strawberry plants should be protected with a layer of clean straw which protects the ripening berries (see Figures 5.13 and 5.14).

This should be put in place as soon as flowers begin forming. Alternatives to straw include the use of inert mats which can be placed on the soil around the neck of the plant. Cleanliness is essential as strawberry fruit are particularly prone to diseases such as grey mould (*Botrytis cinerea*) (see Figure 5.15).

Strawberry beds can become overcrowded as they age (see Figure 5.16). Overcrowding reduces the number, size and quality of fruits so that beds should be thinned during the autumn. Beds also require rigorous weed control. Weeds compete for light, water and nutrients and

Figure 5.11 *Strawberry plant established through black polythene*

Figure 5.12 *Second-year strawberry plant in flower*

Figure 5.13 *Second-year strawberry plants well protected from soil by a layer of straw*

Figure 5.14 *Strawberry fruits kept clean by a layer of straw*

Figure 5.15 *Strawberry fruit showing infection with grey mould* (Botrytis cinerea)*, severe infection, incipient infection and healthy fruit (left to right)*

Figure 5.16 *An overcrowded second-year strawberry bed*

encourage unwanted humidity around the strawberries which increases disease incidence. All weeds should be removed as quickly as possible and certainly before they produce seeds. Perennial weeds with large deeply penetrating root systems such as dandelions (*Taraxacum* spp.), nettles (*Urtica* spp.), docks (*Rumex* spp.) and couch grass (*Elytrigia repens*) are particularly damaging in soft fruit crops if they are allowed to develop. Gently agitate the soil around the plants at intervals after weed removal with a Dutch hoe. This tool is constructed such that soil flows over the blade while it slices through weed seedlings. The most effective weed control is achieved on warm dry days when the weeds will be desiccated. Dutch hoes leave a layer of dry friable soil round the strawberries which helps deter slugs and provides a mulch which conserves soil moisture. Spreading mulches of well-decayed compost also aids the conservation of soil moisture. Mulching is also a means of pest and disease control which is essential throughout the life of strawberry beds.

Strawberries reproduce by natural vegetative propagation, forming runners usually after fruiting. Runners are an elongating stem (stolon) at the end of which a new plant develops (see Figures 5.17 and 5.18).

Further secondary and tertiary runners may form from the primary plant. Consequently, runner growth if left uncontrolled will swamp the bed with small unproductive plants (see Figure 5.16). Remove runners at the point where they emerge from a parent plant by severing with secateurs, a sharp gardener's knife or using the thumbnail and forefinger. Selected vigorous and healthy runners can be retained as a means of generating new plants.

These runners should extend and produce a primary new plant, but not be allowed further growth. Lift the primary runner plant which may already be showing some root initials (see Figure 5.19).

Bury under the runner a pot full of potting compost and peg down the plant with a piece of number 14-gauge wire bent into a "U" shape (see Figure 5.20) and water well (see Figure 5.21).

Very quickly, the runner plant will produce a root system growing into the pot (see Figure 5.22).

Figure 5.17 *Strawberry plant with a primary runner*

Figure 5.18 *Runner plant close up*

Figure 5.19 *Runner plant showing root initials plus pot of compost and pin*

Figure 5.20 *Runner pinned down in the pot of compost*

Figure 5.21 *Water the runner well*

Figure 5.22 *Well-rooted strawberry plant developed from a runner*

At this point, sever the links between parent and runner plant and lift the pot. Place the potted strawberry plant in a greenhouse, in a cold frame or under a glass garden cloche standing on a sheet of polythene which prevents root growth into the soil. Strong and vigorous new plants may be produced by this technique forming replacements in the current strawberry bed or for starting a new one.

Strawberry beds require attention throughout the autumn and winter. The leaves senesce and dry out during autumn, and the gardener should remove the dying foliage revealing the crown. Vigorous removal of old leaves, fruit trusses which may have formed and late produced runners ensures that healthy plants are prepared for the re-growth and fruiting (see Figure 5.23). Truss is the term used for the fruiting structure of strawberries and some other fruit where several flowers are produced collectively from one main stem.

This growth is capable of withstanding winter weather tolerating low or freezing temperatures and snow cover. Inspect the plants regularly, and if they have been eased out of the soil by frost-heave, firm them back into the ground. This is done by gently pushing the crown down into the soil. Draw more soil back around the crown. At the same time, remove unwanted guests such as slugs and snails which are hibernating in and around the strawberry crowns.

As growth re-commences in the spring, new leaves will emerge. Add fertiliser containing potassium (K) and phosphorus (P) but not nitrogen (N). These nutrients encourage root formation and the development of flowers. Adding nitrogen encourages over-vigorous leaf formation, and this growth is susceptible to infection by grey mould (*B. cinerea*) (see Figure 5.15). Flowers develop from the expanding crown which attract insects, particularly bees. These are essential for pollination and successful fruit development. Poor insect pollination results in misshapen, woody, tasteless fruit. Once fruit form, they need protection from birds, slugs and contamination with soil or compost resulting from rain splash. Maintaining clean fruit is achieved by placing fresh straw around the plants and tucking it right under the foliage and up to the crown of each plant (see Figures 5.13 and 5.14).

Figure 5.23 *Tidying a strawberry bed for winter*

Alternatively, mats made from woven bamboo or similar organic materials can be slotted around the neck of each plant and lift the developing fruit above soil level. Protection from rain and birds is achieved by covering strawberry plants with netting or polythene tunnels or glass cloches. Low-level polythene tunnels or glass cloches should allow ventilation so that air flows around the plants and preventing high humidity developing which encourages disease development.

Covered plants require monitoring so that they do not dry out; strawberries lose their lusciousness if deprived of sufficient water. Proprietary mixtures of nutrients can be added into irrigation water and ensures sustained vigorous growth and fruit swelling. Low-level irrigation systems using seep hoses or nozzles are available for use in the garden and have the advantage that they can provide automated watering and feeding regimes. Seep hoses are manufactured with minute holes along their length which allow the slow loss of water into the ground around the plants where it is directly available for the roots. These make for easier gardening during holiday periods.

It is important that water does not remain on the leaves or even worse on the fruit where overhead irrigation is provided. When using a watering can, the flow should be directed away from the crown of the plant. Nutrient mixtures should contain high levels of potassium. Remove blemished fruits as these will provide foci for the development of grey mould (*B. cinerae*) and other pathogens or aphid (*Chaetosiphon fragaefolii*) infestations which will damage both plants and fruit. Proprietary fungicides or insecticides or organic materials should be applied at regular intervals as directed by the manufacturer's instructions.

Once harvesting is finished, remove protective nets or cloches and start tiding the bed. All senescent and dying foliage should be removed. Some primary runners may be cultivated producing a new set of plants.

Routine husbandry care should be provided for established and new strawberry beds throughout the autumn, winter and following spring. Weeds and senescing leaves should be removed and the ground between plants lightly forked over to a depth of 10 cm. This opens up soil which has been walked on and compacted during the harvesting season, returning the soil to a friable and fertile state. Mulch the beds with compost, which builds soil fertility and health extending the life of strawberry beds. These should continue production for 3 or 4 seasons and thereafter beds should be replaced either with newly purchased or home propagated plants. The disadvantage of home-grown plants is that they will accumulate virus diseases which cause a decline in vigour and fruiting. Productivity will decline in any case after about the fourth picking season as a result of viruses, other diseases and pests. Strawberry plants are relatively cheap when bought from garden centres or mail-order suppliers so there is little point in retaining a failing bed. Ensure new beds are planted in a different part of the garden as that avoids replant disorders caused by build-ups of pathogenic soil fungi.

f. Raspberry (woody perennial cane fruit)

Raspberry plants provide very succulent and attractive fruits (see Figure 5.24), particularly when freshly picked on a warm summer afternoon when they give off a very attractive aroma seldom found with purchased fruit.

It is well worth cultivating a row of canes despite the effort involved.

Figure 5.24 *Punnets of succulent raspberry fruit*

Figure 5.25 *Excavating a hole for a supporting post close to established raspberry plants*

All raspberry varieties require a supporting system of posts and wirework for the fruiting canes. These can grow to at least 2 m in length and are blown about by wind, suffering damage, which reduces both yield and quality of the berries. Raspberries produce a substantial weight of canes and foliage so that support should consist of posts of about 2 m length sunk into the ground. This requires a hole about 0.25 m or more deep, depending on the soil type (see Figure 5.25).

Figure 5.26 *Post and concrete for supporting raspberry canes*

The post should be driven into the ground below the hole and filling it with good-sized rocks or rubble and then securing the post firmly with branded quick-setting cement (see Figure 5.26).

Orientating the row in a north to south direction is preferable since this ensures that sunlight passes over it and minimises shading. Stretch 4 or 5 strong steel wires between each pair of posts. Attach the raspberry canes to each wire with 4-ply twine. New raspberry cane roots must be kept moist in a similar manner to those of strawberry plants prior to planting (see Figures 5.27 and 5.28).

If planting is delayed for longer than one day, either place moist compost around the roots in the bucket or "heel-in" the plants into a piece of vacant ground. "Heeling-in" is simply placing plants for a short time in a hole in the garden soil and drawing it back round the roots where they can wait for a few days (maximum) before being properly planted. For this operation, take out a hole which is large enough for the whole bundle of roots. Stand the canes in the hole and cover the root systems with moist friable soil completely surrounding the plants.

Figure 5.27 *Bare-rooted raspberry plants ready for planting*

Figure 5.28 *Raspberry roots protected from drying out while waiting for planting*

This technique allows for delays in planting of several weeks if necessary but lengthy delays should be avoided if at all possible because the plants may commence rooting into the surrounding soil.

In their final position, canes should be planted at 0.5 m distances within the row. Where several rows are planted they should be spaced at about 1.5 to 2.0 m between each line. Raspberries are vigorous plants producing substantial growth which will cause shading between the rows if they are planted too closely. Shoots will emerge from the ground between the rows and should be removed otherwise management of the crop becomes difficult.

Land used for raspberries should have been previously well fertilised with farmyard manure and dug over well ahead of planting. Work the soil into a good tilth with a rake and hoe before planting.

For each plant, take out a hole with a spade (see Figure 5.29). Add multipurpose compost into the hole (see Figure 5.30) and mix the compost and soil from around the hole (see Figure 5.31). Soak the compost and soil with about 5 litres of water (see Figure 5.32). The hole should be sufficiently deep and wide enough that it will comfortably accommodate the entire root system without cramping (see Figure 5.33).

The roots must be kept moist until planted by retaining in a bucket of either water or compost. If the plants have been heeled-in, they should be gently lifted and the roots wrapped in damp old newspaper or hessian sacking prior to planting.

Spread out the roots of each cane around the inside of the hole and sprinkle compost over them. Raise the plant up and down two or three times, ensuring that the roots are well surrounded with compost. Refill the hole with soil, and again raise the plant up and down, further ensuring that its roots are buried under 10 cm of soil (see Figure 5.34).

Once the hole has been completely refilled, tread carefully round the neck of each plant firming it in to the hole and adding more water from a watering can fitted with a coarse rose (see Figure 5.35).

Figure 5.29 *Excavating a hole for planting raspberry transplants*

Figure 5.30 *Adding compost into the hole*

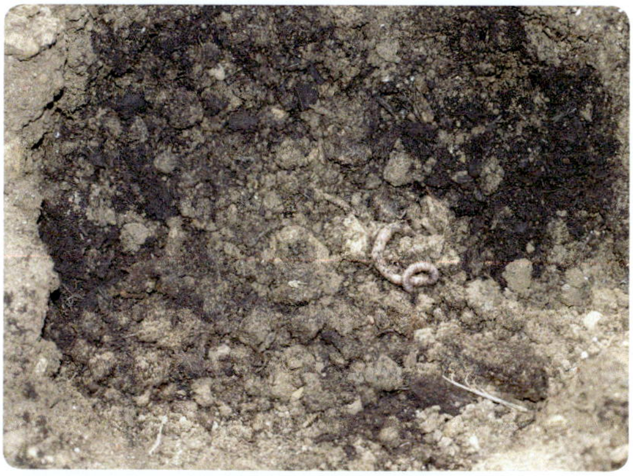

Figure 5.31 *Mixing the compost and soil in the planting hole*

Figure 5.32 *Soaking the compost and soil well before planting*

Finally, add a sprinkling of loose soil on top and repeat this process until all the canes have been planted. As an alternative, use potted plants which may be purchased from a garden centre (see Figure 5.36). These are established in a similar manner to that for bare-rooted plants.

Long cane raspberries should be immediately attached onto the wires preventing wind damage. Working upwards from the bottom wire using 4-ply garden twine tie each cane to the wires. Both long and short canes should be watered after planting and given a dressing of phosphate (P) and potassium (K) fertiliser. This encourages both vigorous root growth and the development of flowering shoots on the long canes. Raspberries should receive very limited amounts of nitrogen (N) as this encourages excessive shoot and leaf growth but not fruit production.

Both long and short canes will form several shoots from the base, which should be trained onto the wires as quickly as possible. These newly forming canes will produce fruit in the following year (see Figure 5.37).

Figure 5.33 *Standing the raspberry transplant in the planting hole*

Figure 5.34 *Covering the root system with soil*

Figure 5.35 *Soak the soil with water*

Figure 5.36 *Potted raspberry plants available in a garden centre*

Figure 5.37 *Raspberry fruit on second-year canes, some ready for picking, others yet to ripen*

All raspberry canes may require insecticide and fungicide treatments preventing damage from pests and diseases. Fruit produced on all canes requires protection with nets or more permanent cages preventing predation by birds, particularly blackbirds. As a temporary measure, drape plastic nets over the canes and pin these to the ground and tie into the wires, making the structure as bird-proof as possible. Erecting fruit cages is a more effective solution but requires considerable planning because an area of the garden becomes permanently devoted for this purpose.

Once the long canes have finished fruiting, remove the nets. As autumn progresses towards winter, the beds may be dressed with farmyard manure (see Figure 5.38).

This is followed by winter pruning. The canes which have fruited should be cut out with a pair of sharp secateurs, leaving a basal stub of 5 to 8 cm length (see Figure 5.39).

Figure 5.38 *Dressing the raspberry bed with farmyard manure in winter*

Figure 5.39 *Removing the canes that have fruited with secateurs*

All pruned canes, fallen leaves and mummified fruit should be removed and disposed of.

A large volume of new canes will be produced by both long and short canes. Select the most vigorous canes and tie these onto the post and wirework. Weaker canes should be removed as these will not form an acceptable crop in the following year.

Normally raspberries produce 5 or 6 good, strong canes per metre run of post and wire work. Limiting the number of canes retained encourages root and fruiting-shoot growth for the following year. Lightly fork over the soil around the retained canes and apply potash (K) and phosphate (P) fertiliser. Apart from bolstering root and shoot growth, adding this nutrient combination in the autumn reduces the likelihood of low-temperature damage during the winter.

During the autumn and winter check that the canes are safely tied into the wires preventing wind damage. Collect leaves as these senesce and fall because they will shelter pests and diseases. Raspberry canes and roots will withstand freezing temperatures and snowfalls.

In spring, the canes will break bud and again and begin forming leaves and flowers. Additional potash (K) and phosphate (P) fertiliser will stimulate further healthy and vigorous growth. Fruiting shoots forming on the canes become obvious in April and May depending on the earliness of the variety selected. Vegetative shoots will emerge from the base of the plants which will form the fruiting canes for the following year. As previously, select the most vigorous and healthy newly forming canes and tie them into the wirework alongside the fruiting ones and remove the rest.

Flowers on the fruiting canes will attract bees and other pollinating insects which should be encouraged. After pollination, fruit starts forming and requires protection. During spells of dry weather in the early summer raspberries need watering at about weekly intervals. All soft fruits contain considerable amounts of water which is the basis of their juiciness and succulence; consequently, drying out reduces fruit quality and quantity. Regularly monitor the plants for signs of pests and diseases. In damp weather, the fruit may become infected with grey mould (*B. cinerea*). Spray with a proprietary fungicide or an organically acceptable alternative immediately if there are any signs of infection following the manufacturer's instructions.

Raspberries should be picked as soon as they ripen. Each berry should detach easily from the central white plug, sometimes a fruit requires a very gentle squeeze, which encourages detachment. Pick over the canes at daily or two daily intervals, ensuring that berries do not become over-ripe and fall to the ground. Check the nets for security from bird predation; this will include releasing birds that have become trapped. Blackbirds in particular are very ingenious at penetrating nets in their quests for ripe fruit.

Once cropping has ceased, redundant old fruiting canes are removed in a similar manner to that used in the establishment year. Where autumn fruiting varieties have been selected, the removal of canes is delayed until late autumn. The pruning routine is maintained for all varieties over several cropping seasons. Raspberries should crop well for 5 to 8 years before their vigour declines and replacements are required. Replacement should be planned well ahead because it takes a couple of years for the establishment of new raspberry beds. These offer opportunities for using different varieties; new ones with increased cropping potential and flavour are emerging very regularly in response to the requirements of the expanding worldwide commercial berry industry. Do not plant a new crop onto land previously used for either strawberries or raspberries. Consequently, fruit cages should be big enough that they will accommodate rotations of cropping over fairly long periods of time. This avoids the incidence of soil borne diseases such as Phytophthora die-back (*Phytophthora fragariae* var. *rubi*) (see Figure 5.40).

Figure 5.40 *Phytophthora die-back (caused by* Phytophthora fragariae *var.* rubi*), a serious root-invading pathogen which is avoided by using officially certified canes*

g. Blackcurrant (perennial woody bush fruit)

Blackcurrant fruits when fully ripe provide an excellent fresh dessert dish and also make superb fillings for summer puddings and jams (see Figure 5.41).

Blackcurrant juice contains high concentrations of vitamin C enhancing human health and wellbeing. Like strawberry and raspberry, blackcurrant plants are supplied as well rooted material usually with several strongly growing shoots which have been formed in the present season or in the previous one (year-old shoots). They are sold as bare-rooted or potted plants obtained from a garden centre or from mail-order suppliers (see Figure 5.42).

Flowers may form in the planting season, this should be discouraged. Pick off developing fruit, encouraging the establishment of a robust plant for future years. Blackcurrant plants are vigorous and will eventually form very large bushes. The roots of "bare-rooted" (unpotted) blackcurrants must, like other soft fruit, be kept moist before planting. Potted plants may be bought from garden centres; these should also be plunged into a bucket of water so that the roots are kept wet. Planting is a priority task; all new plants coming into a garden are safest once they are securely placed in the ground.

Blackcurrants require ample space for each plant, which will easily reach about 2 metres diameter at maturity. Sufficient space should be allocated in advance and well dug and manured during the preceding winter. At planting, take out a hole with a spade deep enough for the entire

Figure 5.41 *Ripe blackcurrant fruit*

Figure 5.42 *Blackcurrant plants in pots on sale in a garden centre*

root system with 10 cm of soil above it. Place several handfuls of proprietary compost into the hole, and fork this into the soil and then soak with a bucketful of water. That ensures the roots are provided with moist, fertile soil. Where appropriate remove the plant's container and tease out the root system with a general-purpose gardener's knife. Cutting into the root ball and breaking it up ensures that the plant can quickly produce vigorous new roots which grow out into the surrounding soil. If the roots are pot-bound and have been trapped in the pot, possibly for several years growing in circles, the plant will not easily form foraging roots. Badly pot-bound plants should not be used. Return them to the garden centre or mail-order supplier with a complaint because such plants will never thrive. They will not form a new root system outside of the existing ball and after a few years will suddenly fail probably following a particularly long spell of dry weather.

Place the root system into the hole, ensuring that there is ample room for individual root development and replace soil around the neck of the plant. Gently shake the plant, settling soil round the roots and firm the whole plant by treading lightly round the bush. Blackcurrant wood is quite fragile and easily snaps during planting, leaving jagged snags, which form foci for diseases. Snags should be removed using a sharp pair of secateurs. Lightly fork over and rake the soil around the plant, removing footprints, and water the bushes well, giving each another bucketful of water after planting. Ensure that during dry periods, more water is provided. Add "Grow-More" general fertiliser after planting at 30 g per square metre. Inspect the plants regularly for pest and disease development. Several species of caterpillars, for example the blackcurrant sawfly (*Nematus olfaciens*), will quite rapidly consume blackcurrant foliage. Consequently, at the first signs of an invasion spray the bushes with a proprietary insecticide or organic equivalent at the rates recommended by the manufacturer. The presence of very swollen buds on the stems indicates infection by Big Bud mite (Gall mite, *Cecidophyopsis ribis*), remove these shoots and destroy them.

Shoots present on the plants when they were bought will grow out during the planting year. These may need trimming especially removing shoots developing in the centre of the bush. Pruning blackcurrant and other bush fruits aims at increasing air movement and reducing humidity throughout the plant; this discourages pest and disease development and increases fruit quality. Blackcurrants flower and fruit on two-year-old wood; this means that shoots formed in the planting year will form fruit in the following season (see Figure 5.43).

The objective of pruning is maintaining a balance between fruiting shoots and those which will follow on in a balanced structure, removing branches that cross over with others or those occupying the centre of the plant. Blackcurrant bushes should not require excessive pruning once they are established.

Growth should be monitored during the autumn and winter removing senescent and dead leaves as they fall. Blackcurrants are deciduous plants losing their foliage in late autumn and then entering a period of dormancy during which they can withstand low and freezing temperatures. Accumulated cold temperatures are a trigger for growth, flowering and fruiting in the following season. As bud growth recommences in spring, gently fork the soil around the bushes and keep monitoring for pest and disease development. If there are several bushes, these should be established at planting through black polythene sheeting, which controls weeds. Without control,

Figure 5.43 *Blackcurrant shoot with developing leaves and flower stalk (strig)*

Figure 5.44 *Developing blackcurrant fruit*

weeds will become an increasing problem as the plantation ages. Where regular late frosts are expected covering the bushes with fleece at flowering time is helpful but does discourage pollinating insects such as bees. This is avoided by laying fleece over the bushes at night and removing it early in the morning

Blackcurrants will form only a modest crop in the second year after planting, which should be picked as soon as the berries ripen. The bushes should be covered with nets when the first berries form colour preventing bird predation. Some recently introduced blackcurrant varieties ripen in unison along the entire strig, which simplifies harvesting into a single picking otherwise individual berries should be removed as they colour (see Figure 5.44). "Strig" is a gardeners' term for the pendulous flowering stalk on which the individual fruits develop and ripen, firstly with the oldest which are those nearest to the main stem.

Pest and disease monitoring should continue after harvesting. Autumnal pruning aims at maintaining the bush structure selectively removing central shoot and branch growth and encouraging air movement. This produces a large spreading bush which will yield considerable quantities of good-quality fruit (see Figure 5.45).

General fertiliser such as "GrowMore" may be sprinkled around the base of each plant at 30 g per square metre. Blackcurrants require regular applications of nutrients. One gardeners' adage suggests that they should be planted "on top of a manure heap" which is possibly rather excessive but does illustrate the nutrient capacities of this crop.

Figure 5.45 *Mature blackcurrant bush*

Figure 5.46 *Rhubarb plants ready for picking*

Incorporate the fertiliser into the soil with a hoe or rake. Remove leaves which fall away and, if the autumn is dry, irrigate with a bucket full of water placed around each bush. Blackcurrant bushes will produce sizeable crops for 10 or more years. But productivity can decline as a result of attacks from virus and fungal diseases and pests. Bushes should be replaced once productivity falls. New bushes should not be planted into land previously occupied by blackcurrants.

h. Alternative soft fruit crops

Rhubarb can also be included with fruit since it requires approximately similar husbandry, although strictly it is a swollen leaf stalk (petiole) (see Figure 5.46).

This is regaining popularity especially early forcing varieties such as 'Champagne'. Older types such as 'Victoria' and 'Stockbridge Arrow' retain popularity for main season crops. This is a very reliable crop which can be forced for early production by covering the crowns with a large flower pot or old bucket.

Gooseberries are another reliable crop which like blackcurrants form large spreading bushes as they mature covered in early spring growth (see Figure 5.47) which has large spines.

Gooseberry bushes should be carefully monitored for the first signs of insect attack, as the plant is vulnerable, the foliage being an attractive food source for sawfly (*Nematus ribesii*) caterpillars (see Figure 5.48).

Attacks can be quite devastating but are controlled with suitable insecticides or alternative organically acceptable remedies. Ripe gooseberries are one of the tastiest soft fruits and quite under-rated (see Figure 5.49).

There are several other soft fruits such as: red currants (see Figure 5.50) and white currants, tayberries, blackberries, a range of other hybrid berries, and also cranberries and blueberries (different species of *Vaccinium*) which are valuable additions for the fruit garden where space permits. Cranberries and blueberries originate from regions with very acidic, peat rich soils and are not therefore suitable for gardens with alkaline pH values. They also suffer from a considerable number of pest and disease problems and are very attractive food for birds.

Figure 5.47 *Gooseberry shoot with developing young leaves and spines*

Figure 5.48 *Gooseberry shoots defoliated by gooseberry sawfly (*Nematus ribesii*) caterpillars*

Figure 5.49 *Ripe red dessert gooseberry fruit*

Figure 5.50 *Ripe red currant fruit*

Underpinning knowledge

5.1 What fits where

Structure is the pattern by which the organs (roots, shoots, leaves and flowers), form the functioning whole which is the plant as determined by its genetic constitution and adapted by evolution fitting the environmental niche which it occupies.

Flowering plants have a basic general structure (morphology and anatomy) consisting of a root and shoot system. The roots provide anchorage and stability, forage for water and nutrients and form beneficial associations with microbial life in the soil (see Chapters 1 and 2). The shoot is a platform for leaves where the functions of respiration, photosynthesis, transpiration and translocation mainly take place. The shoot must also be capable of withstanding the stresses and strains resulting from atmospheric physical forces such as wind, rain and temperatures varying across a range of at least 30 °C. Normally, there is a natural balance between the root and shoot systems of plants. This ensures that supplies of nutrients and water are adequate for the requirements of the shoots and the products of photosynthesis supply sufficient energy for the needs of the roots. Gardeners frequently manipulate this balance encouraging increased fruit and flower formation by pruning and the supply of additional nutrients (see Underpinning knowledge 4.2). Flowers and fruits are ultimately borne on the shoot framework (see Underpinning knowledge 4.1).

Growth is either primary or secondary. Primary growth results from divisions taking place in the cells of the apical meristems resulting in root and stem extension. Secondary growth results from divisions of the cells in the vascular tissues resulting in increasing girth particularly of woody perennial plants such as raspberries and blackcurrants.

As already noted, plants are of two basic types, monocotyledonous (monocot) and dicotyledonous (dicot) types (see Underpinning knowledge 3.2). Essentially, this nomenclature relates to the formation of one or two seedling leaves, respectively. But there are other fundamental differences between these two groups, particularly in the anatomy of stems and roots. In monocotyledonous plants, the vascular tissues are scattered throughout the general parenchyma; while in dicotyledonous types, vascular tissues develop in a solid continuous ring within the general parenchyma (see Figure 5.51: a and b = dicot stem, c = monocot stem and d = comparison of monocot to the right and dicot to the left).

This difference has considerable implications for the structure of garden plants. Moncotyledonous plants such as the grasses, while being very robust and having lifestyles which are capable of living for many years, tend to be relatively low-growing. The structure of dicotyledons permits growth as trees or shrubs.

Monocots form tree-like plants such as palms and bamboos, although their internal structure contains no "wood". By contrast, dicotyledonous types can form giant forest trees and substantial shrubs and also most annual plants (see Figures 5.52, 5.53 and 5.54).

Perennial dicots accumulate considerable woody tissues in the stems and are able to withstand enormous stresses and strains as with those which impinge on very tall forest trees. Cells in the shoot and root meristems become differentiated by function forming into the epidermal, vascular and parenchyma tissues. Cell division in the stem lateral meristems results in secondary growth. These secondary tissues lead on to the formation of cork or bark on the outer edge

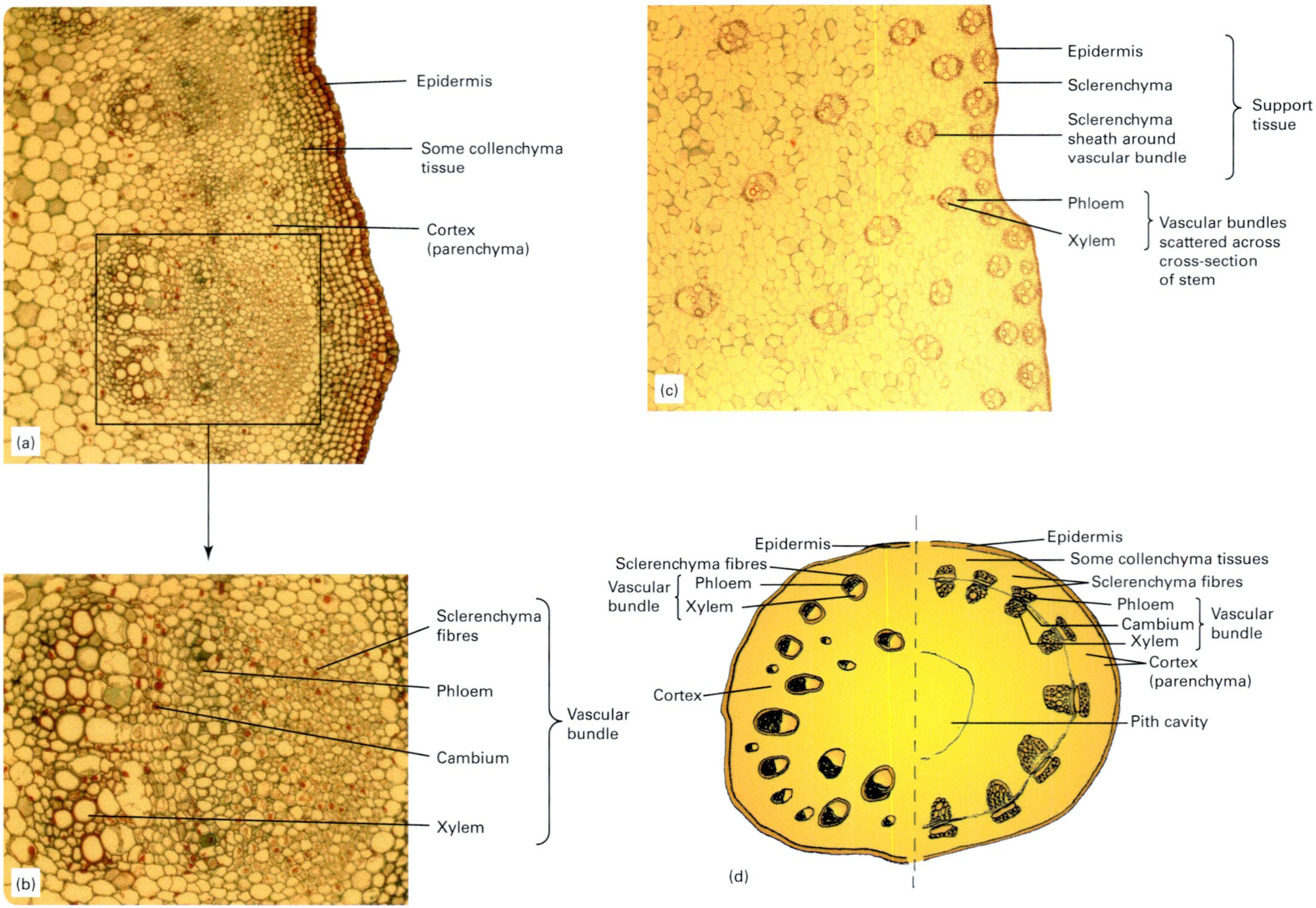

Figure 5.51 *The internal structure of monocotyledon and dicotyledon stems. a and b = typical dicotyledonous stem in section (*Helianthus annuus*, sunflower), c = typical monocotyledonous stem (Zea mays, maize) and d shows a comparison of monocotyledon on the left and dicotyledon on the right*

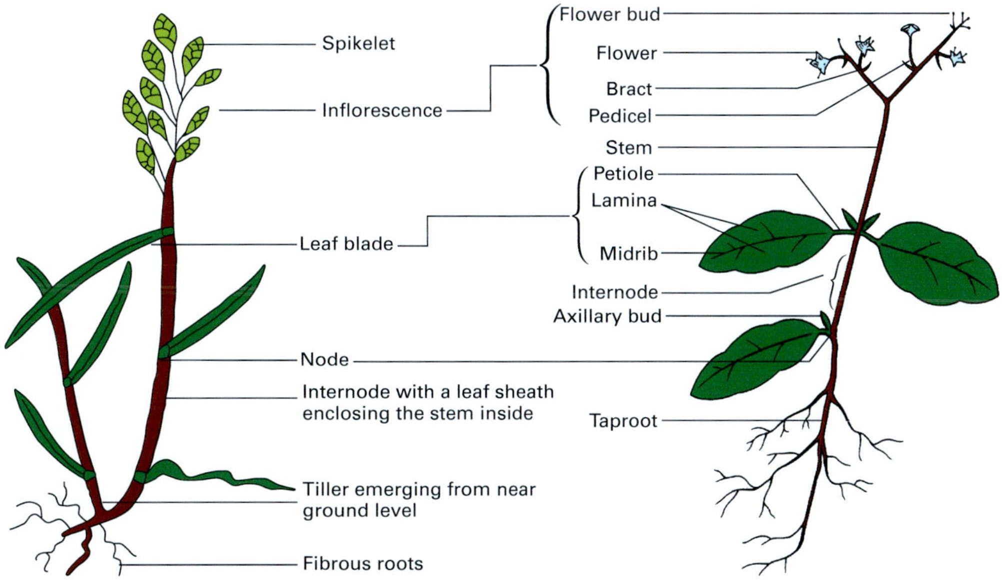

Figure 5.52 *The external structure of (a) monocotyledon and (b) dicotyledon plants*

Figure 5.53 *Dicotyledon leaf,* Bergenia *spp.*

Figure 5.54 *Monocotyledon leaf,* Yucca flaccida *var. 'Golden Sword'*

of the expanding stem. The meristems in the vascular tissues form secondary xylem which is the main component of wood in stems and the secondary phloem on the outer edge. This produces a structure which is essentially a series of concentric cylinders and that results in capabilities for withstanding lateral pressures caused by wind. As a result, trees and shrubs sway and bend in winds of quite considerable forces before they break.

In most herbaceous dicots, the vascular tissues are arranged in rings. Inside the vascular tissue is the central stem pith, while on the outside is a cortex of parenchyma cells and the epidermis. In many herbaceous plants, this outer tissue may be very thin. Woody plants start producing secondary thickening in the vascular and cork cambiums. Differences from herbaceous types are most obvious with the production of added xylem as woody tissue. Growth in cold- and warm-temperate woody plants takes place during the spring and summer. Consequently, trees and shrubs are best planted and transplanted in winter and into the early spring before bursts of growth. Planting before spring growth starts ensures that the plants can take advantage of their new positions as soon as weather conditions are conducive for root growth, usually above 5 °C. Growth is maximised during the most favourable period. Planting and transplanting in summer through to the early autumn is not a wise policy because plants may make little growth until the following spring. That means they are exposed to stresses and strains of wind and rain without having the root structure necessary for stability and anchorage. Planting during winter when these plants are dormant is feasible provided they are well supported by posts and wirework. Wind-rocking of unrooted trees and shrubs leads to the formation of cavities in the soil around the roots, which fill with winter rain, especially where the garden is sited on clay, and this leads to rotting. Plants which have originated from very warm to temperate and tropical climates will grow around the year provided the temperatures are high enough, but in cool to temperate climates fail because of a lack of root growth.

The consequence of growth cessation after mid-summer in woody dicots is the formation of annual rings in the woody xylem vascular tissue (see Figures 4.55 and 4.56). In the early stages of growth in each year, the meristems in the cambium produce large cells. Gradually, as growth begins tailing off, cell size is reduced. This pattern of large and small cells in the woody tissues leads to the formation of annual rings by which the age of a plant and the favourability or otherwise of seasonal growing conditions can be estimated (see Figures 4.55 and 4.56). When a tree or large shrub is cut down, the annual rings are exposed. Rays of cells crossing the rings will be obvious; these are the remains of parenchyma cells which conducted water and nutrients across the stem. The wood is differentiated into central heartwood, which consists of the old and non-functional original cells of the tree or shrub. This heartwood does, however, still provide mechanical strength for the plant. Outside that is the sapwood which functions for the conduction of water and minerals vertically up the stem. Similar structures are found in the peripheral meristem with an outer layer of older bark and an inner layer which still functions.

Monocotyledons lack both the vascular and cork cambiums and their subsequent tissues. Consequently, these plants tend to be herbaceous although they can still attain a considerable age. Stems of monocots resemble dicots by having a surface epidermis and cuticle. But the vascular bundles are scattered through the central tissues formed of parenchymatous cells. There is no cortex and pith structure in the stem of monocots. Since there is no cambium tissue, growth takes place at the nodes where leaves are attached. Cell multiplication at the nodes increases stem length but not girth. Consequently, the stems of monocots tend to have uniform diameter along their entire length. These nodal meristems are important for multiplication by cuttings since roots form from them (see Chapter 8).

Stems are modified in various ways. Each variation is capable of producing roots which are known as adventitious roots (see Underpinning knowledge 8.1), distinguishing them from roots developed from seed. Modified stems offer means for artificial vegetative propagation as:

Bulbs: these are essentially large storage and survival buds with roots coming from a swollen base from which all the leaves develop (see Figure 6.28).

Multiplication is achieved by slicing down through the bulb vertically producing segments of the leaves between which are buds which are dusted with fungicide and can be planted in pots and will reproduce the bulb; examples onions, lilies (see Chapter 8), tulips and daffodils;

Corms: superficially, corms resemble bulbs, but sectioning in half reveals solid tissue that is a modified stem with a few outer paper leaves attached.

Multiplication is best achieved from the small corms formed at the base of the parent: examples *Gladiolus* and crocus (see Figures 6.29, 6.30 and 6.31).

Rhizomes: a horizontal stem which grows either underground or on the soil surface, with leaves growing from the inconspicuous nodes forming axillary buds, adventitious roots (coming from a stem) forming along its length (see Underpinning knowledge 8.1).

Consequently, rhizomes may be propagated by cutting into sections: examples are *Iris* (Chapter 8), perennial grasses and ferns.

Runners and stolons: horizontal stems radiating from a parent plant with plantlets forming at the nodes; some authorities consider that stolons are underground runners; propagation is achieved by cutting sufficiently well-developed plants from along the length of the runner; example, strawberry (see Chapter 5).

Tubers: this term is applied to swollen stems, as in potatoes, where the tips of an underground stolon swell considerably. Artificial multiplication is achieved by cutting the swollen tip retaining the "eye" or axillary bud with embryonic leaves (see Figure 2.18). *Dahlias* are an example of root tubers where a bud is situated at the apex of the root from which a branching aerial network of stems develop (see Figures 6.32, 7.26 and 7.27).

Tendrils: modified stems which grow out from the nodes on a main axis and twine round supports provided by surrounding plants or artificial structures, as with runner beans; some tendrils result from modifications of leaves or petioles; for example, peas (see Figure 5.55), pumpkins and flowering plants such as the passion flower (*Passiflora* spp.).

Cladophylls: modified stems found in some cacti and other succulents; they may be flattened and resemble leaves: examples are various cacti, Butcher's broom (*Ruscus* spp.) and asparagus.

Stems offer a framework for the leaves which are the principle gateway for photosynthesis, respiration and transpiration (Chapter 1). Simple leaves such as those of birch, beech, oak or ivy have undivided structure, although they may have indentations and teeth set round the edges. Compound leaves are segmented into leaflets. Each of these lacks a bud at its base, which contrasts with the whole-leaf examples including walnut and horse chestnut.

Figure 5.55 *Tendrils growing from the leaves of peas*

The leaf has a network of veins. In monocots, these run parallel with the sides of the leaf, while in dicots the drainage pattern is reticulated network (see Figures 5.53 and 5.54). It is through these veins that water and nutrients are transported in the xylem, and manufactured products like carbohydrates are moved in the phloem to a site where they are converted into energy or placed in storage (see Figure 4.61).

Like stems, some leaves have become modified, fulfilling a series of evolutionarily useful functions.

Bracts: these are leaves which develop within the flower adjacent to the sepals, in grasses, they manufacture photosynthates which are used by the developing flower and seeds (grains); alternatively, they have a major floral function, as in *Bougainvillea* spp. and *Euphorbia* spp.

Spines: provide protection for the plant from browsing herbivores and have various shapes from small prickles to very large needles, as in the case of cacti (see Figure 4.60).

Reproductive leaves: some leaves form plantlets around the edge which will detach and root very easily from the end that was attached to the stem, as with *Kalanchöe* spp.

Window leaves: an adaptation for plants which live in desert areas where part of the leaf is devoid of chlorophyll and through which light penetrates into interior cells so that the process can continue even if the leaf is partially covered by blowing sand, for example in *Lithops* spp.

Shade leaves: plants that grow in areas of low light intensity produce large leaves which are capable of photosynthesis spread across the large leaf area. Similarly, plants that grow in tropical rainforests may have pointed leaves which shed water more effectively, an example is the rubber plant (*Ficus elastica*).

Insectivorous leaves: some temperate (e.g., sundew, *Drosera* spp.) or tropical pitcher (*Nepenthes* spp.) plants form leaves that are modified as traps catching insects, which are dissolved by enzymes, providing sources of nitrogen and other nutrients (see Figure 5.56). Some insectivorous leaves can be quite large as with some tropical pitcher plants and capable of capturing rodents.

Figure 5.56 *Insectivorous pitcher plants (*Nepenthes *spp.)*

5.2 Using resources

Resources are cumulatively the environmental materials which permit plant growth and reproduction and the products which are manufactured within plants from them. Interactions between environments and plant growth and reproduction are introduced in Chapter 1. The overall form and function in plants results from the interaction of the genes which are inherited from the parents (known as the genotype) and the environment in which they grow. This interaction results in the visible plant (phenotype). This is known as the genotype x environment interaction which results in phenotype (G × E = P). Recognition of the power of this relationship led Charles Darwin and Alfred Wallace on to an understanding of the forces resulting in the evolution of individual organisms by natural selection (see also Underpinning knowledge 3.1 and 3.3). (Darwin's publication in 1859, *On the Origin of Species by Means of Natural Selection, or the Preservation of Favoured Races in the Struggle for Life*, dramatically changed the scientific basis of biology, horticulture and gardening.)

It is now well accepted that the atmospheric and soil environments powerfully shape the plants discussed in this chapter as woody perennial soft fruit, strawberries, raspberries and blackcurrants. Growth is dependent on the genetic constitution of plants, their DNA, and this is shaped by interdependence on environmental factors such as temperature, moisture, light intensity and day length (see Underpinning knowledge 7.1). The effect of light direction is shown in Figure 3.67 where cress seedlings have been subjected to light from one direction and grown towards it. This is compared with light coming from all directions where the plants grow upwards.

Temperature is a major environmental factor which encourages or delays plant growth provided there is sufficient light, water, nutrients and carbon dioxide. Each of these will be a "limiting factor" if it is present at sub-optimal quantities (see Chapter 1). The soft fruit described in this chapter respond by changing from dormancy into growth once soil and air temperatures consistently rise above 5 °C. Dormancy is a quiescent state in which the buds are protected from low-temperature injury both physiologically and physically. The physiological state of dormant plants is genetically programmed such that it requires an accumulated quantity of cold (accumulated cold units) before growth can take place. Once this value is achieved, the buds are stimulated by chemical messengers, plant growth regulators, into growth and development. Physically, the buds which were formed during late summer and early autumn during the previous growing season developed layers of scales, which is a further protection mechanism. As the buds open, they shed these protective scales and then internal organs expand, forming leaves and flowers. Initially, rising temperature is the major limiting factor, but once the leaves begin expanding then increasing day length is also important (see Underpinning knowledge 7.1). In strawberry, the leaves form a rosette structure, whilst in raspberry and blackcurrant buds on the woody tissues open and begin expanding. These form both photosynthetically active leaves and flower buds. Development starts from specialised meristematic tissues inside the bud.

Vegetative growth also develops in raspberries and blackcurrants in spring. Raspberry plants are stimulated into forming new shoots growing directly from the roots. Traditionally, the raspberry fruit crop ripens in mid-summer (June through to mid-August). But some late-flowering varieties have been bred which ripen in the autumn. Consequently, by growing a spread of varieties, cropping may be extended substantially. Blackcurrants flower and fruit on shoots which are at least in their second season of growth.

Fruiting in each of these crops relies on the pollinating effects of insects, predominantly but not exclusively by honey and bumble bees. This is an example of convergent evolution where the plant and insect have qualities that are mutually advantageous. Plants produce nectar, which supplies energy for the insect. In turn, the insect transfers pollen grains from the anthers to the stigmatic surfaces, which allows the sexual reproduction of the plant (see Underpinning knowledge 3.1 and 3.2). Temperature significantly affects the activities of insects; warmth in the early spring encourages their foraging activities. Consequently, soft fruit pollination can be badly depressed by low temperatures. This becomes obvious in strawberries as the potentially juicy fruits develop only partially when pollination is disrupted by periods of low temperatures. Failed pollination results in some seeds not being fertilised and these do not produce adequate amounts of growth-regulating hormones, resulting in misshapen fruit, often with hard outer surfaces.

Raspberry and blackcurrant plants also require pollination as a precursor for fruit development and ripening. Blackcurrant berries form on a strig of serially developing fruits which may ripen at slightly varying rates particularly in older varieties. The berries at the top of the strig become fully ripe when those lower down are still green. This meant with older varieties of dessert blackcurrants single berries were laboriously picked individually. Plant breeders are overcoming this problem with newer varieties which ripen more uniformly.

After fruiting, late summer and early autumn temperatures must remain sufficiently high, enabling the growth of strawberry runners, and maturation of raspberry and blackcurrant shoots in shortening day length. As temperatures decline through the autumn, these plants acclimate in preparation for dormancy (see Figure 5.57).

Acclimation is a state in which cold tolerance extends and growth ceases. In this process, plants gradually become acclimatised to temperatures which previously would have been damaging. But the process can be reversed if weather conditions change into a warm period, or "Indian summer". In each of these crops, changing temperature and physiology results in a loss of chlorophyll in the leaves and revealing red and yellow pigments as "autumn colours" followed by leaf dehiscence (see also Figure 3.68). Acclimation is a general term developed initially in the United States which describes the initial processes of acclimatisation in autumn leading into the development of winter dormancy.

The formation of an abscission layer composed of callus cells severs the link between the leaf and plant leading to dehiscence. Combinations of changing temperature and day length interact with altered patterns of growth-regulating hormones bringing about an ageing effect. The subsequent onset of very short days and colder weather results in dormancy in all three crops, which are then capable of tolerating temperatures below –10 °C. Unlike acclimation, dormancy cannot be reversed solely by short periods of rising temperatures and or by increasing day length before a threshold is reached with the onset of spring. This is a protection mechanism preventing a resurgence of growth in mid to late winter when it could be severely damaged by a resumption of freezing conditions.

Figure 5.57 *Frosting on a leaf of* Cercis canadensis *var. 'Forest Pansy'*

By comparison, plants from tropical areas continue growing year-round because the temperature does not vary greatly. These hot and humid conditions allow the evolution of plants with large leaves which can harvest maximum light in the lower levels of the canopy (see Figure 5.58).

Alternatively, some plants are capable of existing as epiphytes (see Figure 5.59).

They are able to draw nutrients from the accumulated mosses and lichens which grow on forest trees as a consequence, they live on branches of host trees higher in the canopy where more light is available.

Figure 5.58 *The foliage of plants derived from the wet tropics*

Figure 5.59 *Orchid plants from the wet tropics growing epiphytically*

5.3 Nature's helpers

Pollination is the mechanism by which pollen grains containing male gametes are move to the stigmatic surfaces of the female organs in plants (see Figures 3.58, 3.59 and 3.60). Pollination is vital for many flowering plants and as a result flowers have become key attractions of most plants in gardens (see Underpinning knowledge 3.2 and 3.3). This attribute evolved as a means of attraction aimed at ensuring their sexual reproduction. Flowers are vital for many vegetable and fruit crops as much as for ornamentals. Some species evolved structures allowing the easy movement of pollen in wind or in some cases water. In others, flowers have the vital role of attracting insects that transfer pollen containing the male gametes to the female stigma, ensuring cross-pollination.

Consequently, the evolution of flowering plants is closely linked with that of insects and other animals. Pollinators may transfer pollen between the pistil and stigma within the same flower, or different flowers of one plant or between different plants. This indicates the level of intricacy that has developed in the plant–pollinator relationships. In some instances, this relationship is so precise that only one species of animal can effect pollination.

Flower development

Flowers develop from the apical meristems of shoots which by environmental stimuli, frequently the effects of day length, alter from a vegetative state into one where the reproductive organs are produced (see Underpinning knowledge 3.1 and 7.1). The floral shoots are an example of determinate growth when vegetative cell proliferation ceases and the sexual organs of the flower emerge by expansion of the structure. This contrasts with indeterminate vegetative growth found in some leafy shoots where extension continues by the addition of new cells at the apical meristem. In these plants, flowers develop on lateral shoots.

The structure of flowers has over evolutionary history followed two pathways resulting in the enormous diversity of flowers.

- Firstly, the individual floral parts (sepals, petals, anthers and stigmas) have become grouped together or even fused.
- Secondly, the number of floral parts has been reduced or lost completely. The mathematical arrangement of floral parts has changed from a series of spirals to a single whorl and the central axis of the flower has shortened.

Evolution has also promoted the development of large highly attractive petals. Archaeologically, early flowers had floral structures which were radially symmetrical, so that division in any direction always produced two equal halves. More recently evolved flowers are bilaterally symmetrical and can only be divided equally along a central line. Each of these modifications lead to increased diversity and attractiveness which resulted in greater opportunities for pollination.

Wind pollination

Passive pollination by wind is probably evolutionarily the earliest means of transferring male gametes in wholly land plants, for example in the conifers (Gymnosperms). Wind pollination requires vast quantities of pollen grains which are released from the male flowers and are blown onto the developing female ones (see Figures 3.59 and 3.60). This mechanism requires the very wasteful production of excessive quantities of pollen and restricts delivery to about 100 m.

Examples of wind pollination also include grasses, forest trees such as birch and oak and meadow species like willow. In these species, flowers tend to be small and not very conspicuous, attractive or colourful, possibly green in colour with much-reduced petals and sepals. Frequently, wind-pollinated flowers are pendulous, hanging where wind can most easily gather pollen grains and waft them towards a receptive stigma. Cross pollination, out-breeding, is encouraged in wind-pollinated plants by the male and female organs being separated on different plants. These organs may ripen and be sexually receptive at slightly different times again encouraging out-breeding. In many wind-pollinated types, the flowers mature in early spring before the leaves expand, aiding the passage of pollen unimpeded by foliage. Wind-pollinated tree species may be used in structural planting of landscapes because they lack attractive flowers. Herbaceous wind-pollinated types, especially grasses, add contrast into the garden, acting as foils for floriferous colourful plants. Rushes and sedges add interest for water gardens or elsewhere because of the long pendulous wind-pollinated flower stalks.

Insect pollination

The disadvantages of wind pollination which wastes resources and energy may have driven evolution towards the active carriage of pollen by animals. Insects are one of the oldest and largest groups of animals occurring in all geographical zones and their flying abilities makes them ideal vectors for pollen transfer (see Figure 3.58).

The relationship between insects and flowers started developing more than 100 million years ago. Over that long history, evolution has tended towards increased specialisation between particular insect and flowering plant species. Bees are the biggest group of insect pollinators. Floral characters which attract them include: colour, shape and texture, with blue and yellow colouration and possibly lines and stripes on the calyx or corolla being particularly popular. Some flowers have developed nectaries which is an added inducement for pollination, these exude sugars which bees find very attractive. Bees and other insects identify flowers as sources of food from both the nectar and pollen which feeds both adults and developing larvae. There are huge numbers of bee species which are pollinators mostly these are solitary types, but social bees and especially honey bees (*Apis* spp.), are of major significance. Some bees have modified mouth parts which equip them for a very specialised relationship with particular groups of flowering plants. The work of the Austrian apiarist Karl Ritter von Fisch laid the foundations of understanding of the relationship of bees and flowers. He showed how foraging bees are able to communicate information to other members of the hive as to the location of new sources of pollen. This was accomplished by intricate "dances" performed on the platform outside the hive. Bees recognise flowers by smell and colour and can orientate their location by the position of the sun.

Insects such as butterflies and moths are also active pollinators. Often, they are attracted by white, yellow or other pale-coloured flowers which are flattened, offering landing places. Some of these insects have modified mouth pieces such as elongated proboscises, enabling them to extract nectar from deep inside the flower, while at the same time collecting pollen on their body parts. Proboscis refers to the highly modified mouth parts of some insects which are tubular in shape and enable them to suck fluids from plant tissues.

Bird pollination

Birds, such as hummingbirds (*Trochilidae*), have also adapted as pollinators, usually being attracted by large, very colourful tropical and sub-tropical plants which offer considerable supplies of nectar and pollen. That is an essential element in their relationship since these larger pollinators require that a flower offers considerable sources of energy. Relationships between colour and pollination have resulted in some very intricate relationships. For example, red flowers tend to be pollinated by birds because insects are unable to register this colour. That mechanism ensures that red flowers are reserved for bird pollinators.

Attracting pollinators

Many fruit and vegetable crops rely on pollination for example, edible fruits and legumes. Garden plantings containing a great variety of flower colours and types over the entire season encourages a diversity of pollinators.

Self-pollination

Some plants have by-passed the need for pollen transfer and are solely self-pollinating. These are frequently plants from temperate regions with small insignificant flowers. In some specialised cases, pollination takes place before the flower opens. The arrangement of styles and stigmas is such that the pollen simply drops from one onto the other. Self-pollination is ecologically efficient since there is no reliance on a third party for successful sexual reproduction. That is an important factor for alpine plants. At higher altitudes, insects are only present for short mid-summer periods and these may be out of phase with the reproductive cycles of plants. Early flowering in spring has advantages for the alpine plant because there is ample time for seed development and dispersal. Self-pollination tends towards producing genetically uniform populations. That is also an advantage for in-breeding plants living in specialised environments since it enhances their fitness for a particular ecological niche. When environments change, however, these organisms are at a distinct disadvantage. Evolving increased biodiversity is a long process, and specialised plants lack some of the mechanisms by which it is achieved. Out-breeding, self-pollinating plants produce large numbers of offspring. These are successful weeds exploiting the uniformity of a gardened environment.

In the brassicas, out-breeding fertilisation is achieved because the female organs prevent the growth of pollen tubes from the stigma into the ovary unless it is of the correct compatibility type.

5.4 Shaping crops

The allocation and control of manufacture products, resources, are organisation and management processes by which carbon capture, photosynthesis, and the building of complex carbohydrates, proteins and lipids is geared such that it fulfils needs within the overall organism for current purposes and develops reserves for future uses.

Plants produce their own resources such as carbohydrates, proteins and fats from carbon dioxide, water and nutrients (see Chapters 1 and 2). The ultimate product of these resources is yield as roots, foliage, flowers or fruits. This may be expressed in terms of the stems, leaves and roots which plants produce during vegetative growth. Such production is important for gardeners with crops such as lettuce or cabbage where the leaves are the desired product. Alternatively, the products of sexual reproduction are important constituents of yield in crops such as legumes and soft fruit. In either case, these components of yield require the allocation of resources either for immediate use or stored for future needs.

Perennials like the soft fruits allocate resources for growth and fruiting in the current season and store them over winter for use in the following year. The younger leaves are the main source of resources derived from photosynthesis and these are deployed into organs and tissues known as sinks. This production, distribution and utilisation system is termed a "source-sink relationship" (see Figure 4.61). For example, resources may be immediately used in the roots where their energy is released by respiration for root growth and the intake of water and nutrients from the soil. Alternatively, resources which are not currently needed are 'banked' by the plant in developing storage organs such as a potato tubers or the overwintering rootstocks of strawberry and raspberry plants. Growing and fruiting plants have continuing requirements for the intake of simple substances such as water, nutrients, carbon dioxide and oxygen obtained from the surrounding aerial and soil environments. Regulating these inflows requires energy derived from the products of photosynthesis. Each of these processes accelerates as growth starts or restarts in the spring responding to increasing temperatures and day length. Maximum productivity is reached in the summer, followed by steady run-down leading to the death of annuals and dormancy in biennials and perennials in winter.

Plants have capabilities for regulating the rate at which resources are produced, used or stored. Sinks have a controlling influence on the rate of resource production in the sources. This can be demonstrated by removing potato tubers, the sinks, from growing plants, which reduces the rate of photosynthesis in leaves, the sources. Similar effects are found with fruit such as strawberries, raspberries and blackcurrants. These transport services are illustrated in Figure 1.10.

Development of fruits through to ripening and harvest exerts influences on the leaves resulting in increased photosynthetic output. Once the fruits are harvested, then demand from the sinks reduces and the rate of photosynthesis diminishes. These processes can be observed in the garden. Once raspberry canes have fruited, they die and are removed by the gardener. Strawberry plants replaced the sink formed by ripening fruit by the initiation and development of extending stems and buds, runners or stolons, resulting in asexual reproduction. The gardener controls this process by removing excessive runner growth and concentrating resource application into selected vigorous and healthy new plants.

Gardeners control growth, flowering and fruiting in vegetables, fruit and ornamentals by removing excess foliage, flowers and fruit. This may encourage the development of new leaves, shoots and flowers or concentrate resources into selected fruits which develop enhanced size and quality. Sterility in fruit where seed development is absent allows the production of larger and more succulent fruit and its use in the garden is a fairly recent example of the artificial manipulation of plant biology. Gardening could be defined therefore, as "the informed manipulation of source and sink relationships in cultivated plants". A particularly attractive example of this definition is found in the flower garden. Cutting flowers from, for example, roses, sweet peas, *Chrysanthemums* or *Dahlias*, encourages further bud development and more luxuriant blossoming.

The source to sink relationship is a biological example of supply and demand processes. The term "supply and demand" was first used by James Denham-Steuart in 1767 in his *Inquiry into the Principles of Political Economy*; it is now appreciated as applying also to biological processes. The transport system serving this relationship is the vascular tissue which moves resources in the phloem cells from leaves either to the roots or into developing flowers and fruits. Resources can be transported over considerable distances, as with forest trees where the roots require energy supplies for continuous water and nutrient intake. Gardening, however, usually arranges that supplies satisfy the biggest local needs first. Flowers or fruit will be serviced by locally placed leaves. For example, in crops like dessert grapes summer pruning ensures that at least three leaves are retained adjacent to each bunch of fruit. Gardeners enhance the efficiency of the sinks by thinning out the number of berries or other fruit, ensuring that those which are retained develop succulence and flavour. Grapes are very strong sinks and only after harvest are resources diverted to the roots.

The environment affects the effectiveness of source and sink partitioning of resources. Drought, shortage of nutrients or high temperatures affect the productivity of photosynthesis, reducing resource supply for storage and eventually for current growth. Sinks in woody plants result in increasing girth as well as stem length. Consequently, the size and structure of annual rings are altered by favourable and adverse environments and the resultant supplies of resources. One of the largest physiological changes in plants is that from the vegetative to reproductive state, which results from alterations in day length or other stimuli. Flowers and ultimately fruit are demanding new sinks that require substantial resources.

Plants have evolved considerable flexibility in response towards changing environments. Variations in temperature between day and night and between root and shoot alter the fate of resources. If conditions at the root are less favourable for storage, then resources may be diverted elsewhere in the plant. These responses are controlled by an array of plant growth-regulating hormones. Understanding of their functions, interactions, activation and de-activation in responding to environmental conditions is slowly being developed. Ultimately, this research will result in varieties which are more attractive and fruitful for the garden.

Achievements: have you understood and learnt?

At the end of the chapter, you should be able to:

1. Describe the differences between types of soft fruit plant such as: strawberry, raspberry and blackcurrant, their suitability for particular environments and the importance of variety selection.
2. Know how soft-fruit plants should be handled on receipt from a garden centre or mail-order company and how land is prepared.
3. Know the steps required for the preparation of a strawberry bed for its first fruiting season, especially in relation to the exclusion of weeds, pests and pathogens, and protection from predation by birds and slugs.
4. Know how a strawberry bed that has been cropped is treated after fruiting and through the autumn, winter and into subsequent fruiting seasons.
5. Understand the differing types of raspberry cane and their treatment for fruiting in the year of planting or delaying fruiting until the year after planting.
6. Know how a support system for raspberry canes may be established in the garden.
7. Know the cultural requirements for raspberry canes through the autumn, winter and spring periods.
8. Describe the differences in frost-tolerance between differing blackcurrant varieties which alter their suitability for use in varying parts of the country.
9. Understand the reasons for limiting fruiting by blackcurrant bushes in the year of planting, and how blackcurrant bushes should be treated in the autumn, winter and spring and in preparation for the first fruiting and subsequent seasons.
10. Describe other forms of soft-fruit which may be valuable additions to the garden.
11. Describe the differing growth patterns of herbaceous and woody perennials and the anatomical mechanisms of growth which govern stem extension and expansion.
12. Describe the structure of herbaceous and woody dicotyledonous stems and the effects of environmental conditions on the manner by which cells of differing sizes are formed in rings within stems.

13. Understand the differing types of stem and leaf modification which have developed for the perennation of plants over several seasons and the importance for gardeners when these are used for propagation.
14. Describe how the environment and genetic constituents interact affecting the expression of characteristics in garden plants and the relationships between acclimation and dormancy.
15. Understand the importance of flower structure for pollination and the differing forms of pollination mechanism and the differences between wind, insect and other agencies in bringing about pollination.
16. Describe the source and sink relationship in relation to the production and use of resources within plants and that resources are produced by photosynthesis largely in leaves and are transported by the vascular system to sinks where they will be utilised.
17. Know the importance of mineral and water intake through the roots as part of the production of resources by plants.
18. Describe the meaning of the term "components of yield".

Definition of terms: *Describe* is able to show why a scientific principle is important; *Understand* applies scientific principles into the context of gardening; *Know* carries through scientific principles in practical gardening activities.

Chapter 6
Bulbous plants

Introduction

Bulbs are one of the best investments for flower gardens. They require relatively little outlay, either financially or in terms of work, and yet provide excellent colourful displays from mid-winter through early summer and even into autumn. Some bulbous types will provide continuing displays for subsequent years, spreading and propagating into what becomes an expanding and stunning display. Naturalising bulb plantings, for example, amply repay the original investment. Alternatively, they may be planted in containers (see Figure 6.1).

Figure 6.1 *Tulips flowering (multi-flowered var. 'Flaming Club') in an ornamental urn*

Bulbs also offer great visual value when planted into pots, tubs and urns, embellishing hard landscaping features such as patios and piazzas.

The term "bulbous" is used here as a generalisation covering several types of overwintering and dormant reproductive structures including corms and rhizomes. Once planted, these establish and continue growing, forming substantial components of a flower garden. Biologically, they resemble the asexually reproduced structures already described and used in the vegetable and fruit gardens (Chapters 2 and 5). Characteristically, they are large pieces of vegetative

Figure 6.2 *Massed tulips in a border*

Figure 6.3 *Massed early-flowering daffodils,* Narcissus cyclamineus *var. 'February Gold'*

Figure 6.4 Scilla peruviana, *Portuguese squill*

tissue which have the capacity for independent growth and reproduction and multiply forming progeny asexually and with increasing impact and value. For example, forms of daffodil (*Narcissus* species) when naturalised embellish lawns and other features creating long-term garden features. Some bulbs by their very nature provide regular early yearly displays in borders and when mixed in with herbaceous plants, shrubs and trees offer colourful combinations with spring and summer growth and other forms of flowering (see Figures 6.2, 6.3 and 6.4).

Garden practices

a. Flowering sequences

Whilst bulbs are capable of producing stunning displays when in full flower these periods are individually relatively short, normally 10 to 20 days, depending on prevailing weather conditions and the husbandry care they receive. A sequence of bulbs with consecutive flowering periods will, however, extend displays for several months. Achieving flowering sequences requires studying bulb catalogues and horticultural publications during mid to late summer and perhaps visiting local and national flower shows where information is available from bulb companies and nurserymen. Shows include the Royal Horticultural Society's (RHS) events staged around the country, Gardening Scotland in Edinburgh organised by the Royal Caledonian Horticultural Society, plus the Harrogate and Malvern Shows arranged independently.

There are literally hundreds of bulb types and varieties so that initially the gardener is best served by selecting some of the older and well-known types. Some bulbs, such as daffodils and tulips, are divided into differing categories relating to their flower form. Here again, the gardener should select as a start some of the more conventional flowering types.

b. Long-term bulb displays

Successful planting of bulbs for long-term displays requires types which are robust and capable of rapid re-propagation. Very traditional daffodil varieties such as 'Carlton' and 'Sempre Avanti' are ideal for this purpose and are used commercially because of their rapidity of self-reproduction and volume of flowers produced. Most tulips should be avoided for this purpose because they reproduce very slowly, rarely making vigorous displays, but some tulip species which are close to wild types can be used successfully. Buying bags of mixed daffodil varieties is good value for money because they provide sequences of flowering over several weeks and contain types capable of self-reproduction. Single varieties display well in pots and urns, and here tulips such as forms of "Apeldoorn" develop uniformly, producing exuberant splashes of colour, albeit for limited periods. The "species" bulbs, corms and rhizomes establish and grow vigorously providing increasingly colourful displays over several years. This term "species", in a garden sense, describes plants that have not been heavily selected by plant breeders and still closely resemble the original wild types which grow naturally and have been brought into cultivation. Biologically, the term "species" describes a group of organisms that have closely similar characteristics and will interbreed with each other (as described in Chapter 3).

The flowers and foliage of most bulbous plants senesce and turn papery brown some weeks after flowering and then the bulbs enter a period of extended dormancy. This repeats the natural behaviour of bulbous plants, which become dormant within the soil during the hottest parts of years mostly in Mediterranean and other warm temperate regions. In gardens, that behaviour can leave borders looking bare and disappointing, particularly during the summer months. Removing bulbs such as daffodils and tulips after flowering, and replacing them with summer bedding, is one solution that ensures continuity of colour and interest. Alternatively, the bulbs are planted within an herbaceous border with perennial plants and these takes over the task of providing coloured interest throughout the summer and early autumn. Garden planning is very important since herbaceous borders and summer bedding schemes demand considerable organisation and effort.

Bulbs which are dug up after flowering can be replanted in reserve areas of the garden and will eventually reproduce and form flowering-size bulbs in 2 to 3 years. In small gardens, this is rarely worthwhile and probably the best policy is discarding and repurchasing new and possibly different varieties for the following season. This provides variation in colour and form year-on-year which adds to the interest of gardening.

Bulbs used for naturalising and the small species types should be left *in situ* where their populations will expand for future years. This is particularly the case for corms such as crocus and the rhizomatous irises. Bulbs that are left in the ground will require supplementary nutrition. Apply fertiliser as regrowth commences, normally in the late summer or early autumn. Bulbous types begin rejuvenating the root systems well ahead of the appearance of foliage above ground.

A general fertiliser such as "GrowMore" applied at 30 g per square metre is sufficient. Like other asexually reproduced plants such as potatoes, onions and soft fruit, pests and diseases are an ever-present threat for bulbous plants because of their longevity, which allows build-ups of invaders. Virus diseases in particular accumulate reducing the vigour and colourfulness of these plants. Ultimately, infections may become so severe that removal and replacement is the only option.

c. Bulbs "in-the-green"

Purchasing bulbs "in the green" in the early spring, particularly snowdrops and crocus corms increases the rate of establishment and subsequent successful flowering. This technique is particularly rewarding for those small bulbs which establish erratically and with difficulty from dried bulbs and corms. On arrival treat the plants in a similar manner as for soft-fruit plants described in Chapter 5.

The priority requirement is keeping the roots moist; if these dry out then establishment becomes much less successful. Unpack the plants and plunge their roots directly into a bucket of water level with where the foliage emerges from the bulb or corm. As soon as possible, plant the bulbs; their safest home is in the ground. Take out a planting hole with a hand trowel and add some multipurpose compost (see Figure 6.5).

Arrange the roots in the planting hole. Often these bulbs come as bunches and planting a group in one place increases their impact when flowering in the following season (see Figure 6.6). Cover the roots with soil and compost (see Figure 6.7) and water well. The plants

Figure 6.5 *Planting hole for "in-the-green" snowdrops*

Figure 6.6 *Snowdrop bulbs "in-the-green" planted in the hole*

Figure 6.7 *Snowdrop bulbs covered with soil*

Figure 6.8 *Snowdrop bulbs recovered from planting after 10 days*

have considerable capacities for recovery and within a few days will have formed new roots with the foliage becoming erect and possibly even the odd flower being produced (see Figure 6.8)

There will be little display in the year of planting, and any flower heads or seed pods should be removed so that all energy is directed towards establishing a vigorous root system. The foliage will die down and senesce, leaving little or no trace of the plants until the following year; consequently, mark their positions with short bamboo canes, ensuring that they are not disturbed. The display provided after 12 months should be a welcome harbinger of spring (see Figure 6.9). For relatively little outlay, "in-the-green" plants will provide colourful and expanding displays, especially with very early flowering types.

Figure 6.9 *Snowdrops in flower 12 months after planting "in-the-green"*

d. Growing bulbs and similar plants

Bulbs are bought from garden centres, other retail suppliers, mail-order companies and some specialist bulb growers usually in the early autumn. Some of the latter are very specialised offering botanically rare and exotic plants which appeal to well-established gardeners who have developed particular interests and expertise. The more general suppliers are probably most suited for basic gardening interests. As with other vegetative asexually produced materials such as soft fruit, bulbous plants should be accompanied by information that identifies their source and, in some instances, certifies their "virus-tested" status which ensures they will produce vigorous, healthy plants and flowers.

Figure 6.10 *Tulip bulbs in a net bag as received from a supplier*

Bulbs will arrive in paper envelopes or small plastic netting sacks from the mail-order companies and probably in similar containers from a garden centre (see Figure 6.10).

Open the packets and store the bulbs in a cool, darkened space such as a garage, ensuring that the bulbs are well protected from predation by vermin such as mice. Bulbs contain sugars, amongst other chemicals, and vermin will quickly sense their presence. Avoid placing bulbs near sources of heat, as this will injure the flower stalk that is developing within them. Bulbs can withstand low temperatures but not freezing ones. Planting most bulbs takes place in early autumn as soon as conditions permit; as with other plants, the sooner bulbs are placed in the garden, the better (see Figure 6.11).

Flower borders are prepared in much the same way as vegetable and fruit gardens. Where the garden is part of a newly built property, the land will require digging-over and the addition of animal manure or well-rotted compost before any planting can start. This is the work of the first year of occupation, the land should then lay fallow for the subsequent winter and be re-worked in the following spring and summer, removing all weeds as they emerge. Detailed planning and planting commences in the following autumn. This extended period without providing colour or reward from the garden will eventually repay substantial dividends for the careful gardener by the absence of perennial weeds and the friable, fertile, healthy status of the soil.

Established gardens should be worked in line with their previous uses. Although the land may also benefit from applications of well-rotted compost or farmyard manure building up fertile and friable border soils. When purchasing an established garden, the new owner may feel that the layout requires change. As a result, new borders will be created, and these require management which increases their fertility.

Figure 6.11 *Daffodil bulbs prior to planting, mixed trumpet types*

e. Planting bulbs

Examine each bulb for any signs of damage or disease; only use good, sound bulbs. Quality daffodils (*Narcissus*), for example, will probably be double-nosed, indicating the likelihood of a substantial display (see Figure 6.12). Bulbs placed in containers require relatively little work and can be arranged on a bed of multipurpose compost (see Figure 6.13).

Figure 6.12 *Robust double-nosed daffodil bulb*

Figure 6.13 *Tulip bulbs var. 'Apeldoorn Elite' planted in a tub*

Figure 6.14 *Planting daffodil (var. 'Elizabeth Ann') bulbs in a border*

Arrange the bulbs in a dense arrangement, as this will aid their abilities for mutual support in the following spring, and cover with 8 to 10 cm of compost.

For planting in borders or naturalising, use a trowel or long-armed bulb planter and take out a hole where each individual bulb will be planted. Place a layer of 2 to 3 cm of horticultural sand at the bottom of each hole (see Figure 6.14).

Sand aids water drainage around the bulb especially if the soil contains large amounts of clay. This can become waterlogged which encourages bulb-rotting diseases. Place the upright bulb carefully in the hole and add more sand and compost on top and replace as much of the original soil as possible. Lightly firm the soil with your fingers but avoid heavy pressure. Continue planting until all the bulbs have been dealt with allowing 5 to 10 cm between each bulb, depending on their size. Small species type bulbs may be planted at 2 to 5 cm distances between bulbs. Displays from these types of bulb have greater impact where they are massed together. All bulbs must be handled carefully because they contain a flower shoot which is easily damaged or distorted. Daffodil bulbs are relatively robust, but tulips in particular are easily injured. The main reasons why bulbs fail are: they are planted too deeply; the soil is very wet, poorly drained and heavily compacted; the bulbs are of poor quality, possibly affected by diseases; weed competition on emergence of the plants; and the effects of vermin such as mice and squirrels.

The planting process requires care, particularly with daffodils which are being naturalised in a grass sward, as each planting station will be occupied by the bulb and its progeny for several seasons. Each hole should be sufficiently deep so that the bulb is buried to a depth that is equivalent to twice its length. For example, a 2 cm long bulb should have 4 cm of soil above it. Before planting, examine each bulb carefully, ensuring that it is healthy and the basal plate from which previous root growth emerged is placed at the bottom of the hole. Bulbs planted upside down are capable of producing a shoot which will sense the correct direction for growth and reaching the soil surface. But this increases the time from shoot emergence to new foliage developing, weakening the bulb and making the floral display erratic.

f. Filling containers

Containers planted with bulbs such as daffodils, tulips or hyacinths provide excellent early colour in the garden. Containers may be plastic, wood or clay (terracotta) (see Figure 6.15).

Plastic pots are satisfactory bulb containers but the wooden or clay types add elegance to the display. In either case, place a layer of broken pot shards in the bottom of each container. These ensure adequate drainage and effective aeration in the compost. Containers which lack holes in the base should be avoided as water will accumulate in the compost, resulting in waterlogging and anaerobic conditions that damage bulb growth. The concave surfaces of pot shards should face downwards, ensuring that there are gaps around them through which water can escape. The shards also add weight to the container, making it more stable and avoiding wind rock. Fill the container with a suitable proprietary compost, leaving about 5 cm between the surface of the compost and the top of the container. This allows for easier watering once the container is planted.

The bulbs may be arranged as suggested in Figure 6.13. Handle all bulbs with care, particularly tulips which are very easily damaged, causing distortion of the emerging flower. Watering is essential for bulbs grown in containers; rain may wet the surface of the compost but very rarely provides the volumes of water needed. Make sure that the containers are flooded at regular intervals so that the entire volume of compost is kept moist. Compost can dry out, especially round the edges of containers, in which case water will simply run out through the gap round the periphery, depriving the bulbs of sufficient moisture. Re-wetting compost which has dried out is extremely difficult. The only satisfactory way of rewetting compost in a container which has dried out is placing the whole vessel in a much larger one which is filled with water and soaking for a prolonged period. Compost which contains moisture-retentive granules helps ensure that the bulbs have an adequate

Figure 6.15 *Forced ("prepared") hyacinth bulbs for winter flowering indoors*

supply of water. Placing a 2 cm deep layer of pea-gravel as a mulch on top of the compost also increases moisture retention. Do not feed bulbs in containers since that will encourage excessive leaf growth and probably diminish flowering. As leaves and flowers grow, support may be needed, especially for large-headed varieties. Use split bamboo canes and 4-ply twine, providing at least enough support for the entire length of the flower stalks as they extend. Several rounds of string placed at different heights may be needed as the combined weight of leaves and flowers can be quite considerable, and this ensures that the flowers are not damaged by wind.

g. Flowering

The leaves and flowering shoots of many bulbs begin emerging in early spring. These should not be damaged by cultivation or by footfalls when for example, planted for naturalisation into grass (see Underpinning knowledge 6.2). Bulbs planted in containers, borders or naturalised will individually provide colour for two or possible three weeks depending on the prevailing weather. Cool conditions lengthen the flowering period while warm days shorten it.

Once flowering is completed, remove the flower heads and seed-pods, but retain the leaves without disturbance for at least 8 weeks. Leaves should not be removed prematurely before natural senescence is completed, as during this period photosynthesis restores energy reserves which are essential for flowering in the following season. As described in Chapter 5, flowers are sinks for the products of photosynthesis and once their functions are completed the bulb becomes a priority sink for the storage of reserves. Only when the leaves have senesced naturally may they be removed. All residual dried foliage should be raked off the border and the soil surface very lightly forked over, ensuring that the bulbs are not injured.

h. Post-flowering

Bulbs should be removed if they are not being retained for a further season. This applies particularly for bulbs grown in containers. Retaining bulbs from containers for future years is a perennial temptation. But they will not produce satisfactory crops of flowers for several years because the nutrient reserves are much reduced and the bulbs may have divided asexually producing large quantities of small bulbs which are not of flowering size.

Daffodils or other bulbs such as hyacinths which are retained require little by way of pest and disease management. The foliage of tulips, on the other hand, can become infected with tulip fire (*Botrytis tulipae*) disease (see Figure 6.16).

This fungal disease destroys the foliage, making it even less likely that a satisfactory display of flowers will be achieved in the following season.

Figure 6.16 *Tulip fire (*Botrytis tulipae*), one healthy and two infected leaves*

i. Autumn work

In the early autumn, work over ground where bulbs have been retained very lightly with a fork. Carefully ensure that the tines penetrate no deeper than 2 or 3 cm, as the bulbs will probably already have started into growth. Spread "GrowMore" general fertiliser at 30 g per square metre, this provides readily available nutrients for root and shoot growth. Naturalised daffodils and some other bulbs propagate asexually over many years, increasing the annual display. In some seasons, flowering will be sparse. This indicates that the bulbs have multiplied asexually and the progeny have not yet reached flowering size. If, however, sparse flowering continues for several seasons, remove all the bulbs and replant with new stock. Persistent sparse growth can indicate that the stock may have become infested with viruses or eelworms, both of which destroy the flowering capacity permanently.

j. Specialist bulbous plants

An enormous range of bulbous plants is available and these will provide colour and interest round the year. Husbandry treatments as already outlined are suitable for most of them. Cyclamen provide one of the earliest and most spectacular displays and if left undisturbed over years will produce a spreading mass of colour (see Figure 6.17). A complete contrast in garden design is achieved later in the season by growing lilies (see Figure 6.18). These stately plants when grouped *en masse* provide a stunning display which also emits wonderful scented aromas especially on warm evenings adding considerably to the joys of gardening.

Hyacinths, which flower in mid to late spring, also produce well-scented displays (see Figure 6.19). They will also colonise borders, providing return flowering over a number of seasons and are available in red, white and blue colours, offering opportunities for interesting bedding schemes.

One of the garden's least demanding and most rewarding bulbous plants is *Iris unguicularis* (*stylosa*) which thrives on poor, nutrient-starved soil in the hottest and driest parts of the garden (see Figures 6.20 and 6.21).

Figure 6.17 Cyclamen hederifolium *in flower* en masse*: spring flowering*

Figure 6.18 Lilium *in flower: summer flowering*

Figure 6.19 Hyacinthus orientalis *var. 'Royal Navy' in flower: spring flowering*

Figure 6.20 Iris unguicularis (stylosa) *in flower in a border: early spring flowering*

Figure 6.21 *Close-up of* Iris unguicularis (stylosa) *flower: early spring flowering*

Flowering by this plant seems to have been advanced into late December from its previous season of late January and early February as a result of climate change. Irises as a group provide wonderful garden displays and none more so than the "flag irises", which vary in height from 1 m to about 40 cm, allowing for the creation of a border filled with these stately plants. These also require hot, dry locations where they will spread, producing multicoloured displays (see Figures 6.22 and 6.23).

All these plants require minimal husbandry, but should be dug up and re-propagated every few years (see Figures 8.23 to 8.28). Similar treatment is required by *Agapanthus* plants. These are well suited for decorating into gravel driveways and are available in a range of colours (see Figure 6.24). Repropagation is illustrated in Figures 6.35 to 6.39.

Careful considerations are required for specialist types such as those that are autumn-flowering like the colchicums. Planting autumn-flowering colchicums using dry bulbs requires that they are bought in the June to early July period when they are dormant or growth is just starting. These plants will then flower in September, October and November (see Figure 6.25) and die down as the flower stalks senesce; no foliage is produced at this time. In the following

Figure 6.22 Iris germanica *in flower, single colour for falls and upright petals: summer flowering*

Figure 6.23 Iris germanica *bicolour in flower, different coloured falls and upright petals: summer flowering*

Figure 6.24 Agapanthus *spp. in flower: summer flowering*

Figure 6.25 Colchicum autumnale *in flower: autumn flowering*

spring, copious foliage appears from the colchicum bulbs which should be grown unhindered, allowing continuing photosynthesis for as long as possible. This restores the viability of the bulbs and provides resources which will be utilised for flowering in the following autumn. These plants will provide foci of great enjoyment at a time of year when there are fewer flowering plants in the garden. Once established, autumn-flowering colchicums will multiply, producing carpets of colour in September and October. Also capable of providing a colourful autumn display are the Guernsey lily (*Nerine bowdenii*) (see Figure 6.26). The common name Guernsey lily is also applied to *N. sarniensis* which produces deeper red flowers. *Sternbergia lutea* is also a valuable autumn flowering bulb producing masses of deep yellow flowers in the autumn. Foliage appears in the following spring. Small, dry bulbs of *S. lutea* are available from garden centres and by mail order. These should be planted into pots of compost and nurtured in an unheated greenhouse where they will produce foliage. Plant them "in-the-green" into a border where they will become established and provide displays alongside the autumn colchicums.

Figure 6.26 Nerine bowdenii *('Guernsey Lily') in flower: autumn flowering*

Underpinning knowledge

6.1 What are "bulbs"?

Bulbs in the widest sense are organs which store resources through seasons which are too hot and dry for successful growth. Plants forming bulbs, corms and tubers are known as "geophytes", literally "earth" (*geo*) "plants" (*phyton*). These are storage organs providing the plants with capabilities for survival during environmentally adverse conditions such as periods of high temperatures or extreme water shortage. In these periods, bulbous plants enter dormancy, and some have very strong roots which contract, pulling the storage organ further downwards into the soil profile, thereby increasing abilities for survival. Once the temperatures fall and water supplies are restored, vigorous feeder-root growth re-commences, followed by the development of shoots, leaves, flowers and seeds.

The term "flower bulbs" is often used colloquially covering true bulbs and also corms, tuberous and rhizomatous plants. These are, however, botanically distinct structures with similar primary functions for the storage of resources, nutrients and water. This is a diverse plant grouping based on function spanning the large taxonomic divisions of monocotyledons and a few dicotyledons (see Figures 5.51 to 5.54). Preponderantly, these plants are monocots but some dicots have evolved along similar pathways.

Figure 6.27 *Daffodil bulb in longitudinal section*

Bulbous structures

Sectioning true bulbs longitudinally exposes the scales, which are leaves developing from a much-reduced stem known as the "basal plate" (see Figures 6.27 and 6.28).

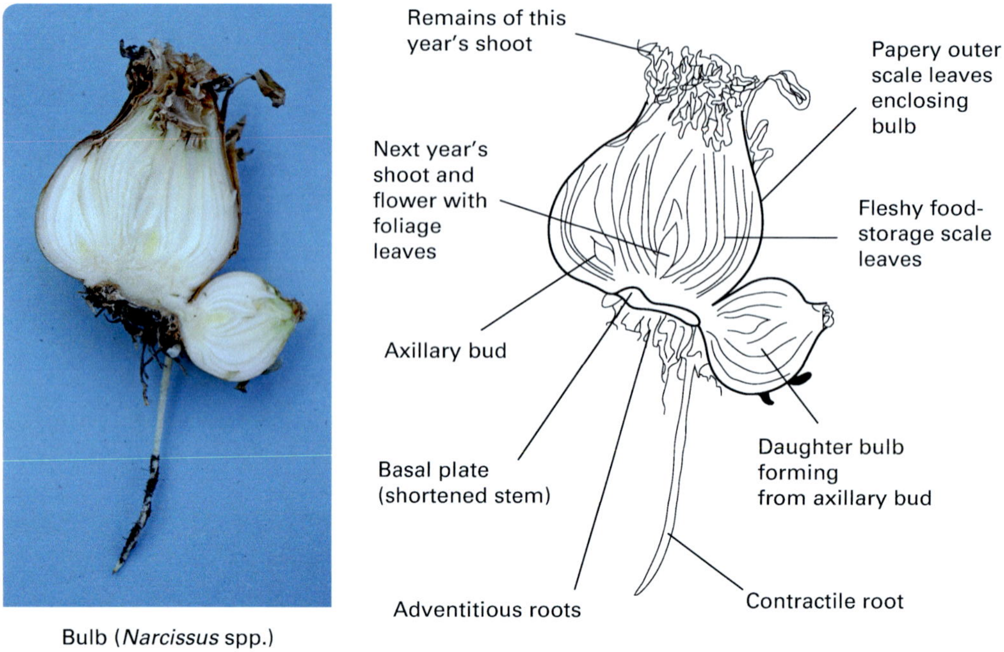

Figure 6.28 *Diagrammatic section of a daffodil bulb (*Narcissus*)*

Also developing from the basal plate are roots which since they grow directly from a stem are known as "adventitious roots". This distinguishes them from roots that grow from an embryonic radicle (see Figure 3.33). Fleshy scales which will appear as leaves when they emerge through the soil surface compose the majority of the bulb. In the centre of these scales is an expanding flower stalk, which may eventually form a large flowering spike carrying many individual blooms or one single blossom. Wrapping round some bulbs is a dry protective tunic, as in tulips and narcissi. This is absent in others such as lilies (see Figures 8.29 to 8.32) and fritillarias. Bulbs reproduce asexually by forming bulblets which grow from vegetative buds present at the base of the scale leaves, often developing around the periphery of the basal plate.

True corms are monocotyledonous plants, for example crocus or *Gladiolus*, and they form from much enlarged stems or basal plates; on this structure are distinct nodes and internodes visible to the naked eye (see Figures 6.29). The whole corm structure may be covered in several dry leaves developing as an outer tunic as found with bulbs (see Figure 6.30). Adventitious roots form round the basal plate (see Figure 6.31).

Corms are, like bulbs, enlarged storage organs. Perennation and asexual reproduction are achieved by annually replacing the current corm with a newly formed one developing on the upper surface of the original parent. Small corms (cormels) also develop around the periphery of the basal plate. Flowers are initiated in the central tissues of the corm with new buds developing around the periphery of the floral stalk.

Figure 6.29 Gladiolus *corm in section, showing last and current seasons' corms*

Figure 6.30 Gladiolus *corm with scale leaves removed, showing buds*

Figure 6.31 Gladiolus *corms showing root emergence*

Figure 6.32 Dahlia *tubers*

The term "tuberous plant" includes several other distinct botanical storage structures. True tubers are enlarged stems carrying one or more buds which grow into stems, leaves and flowers. Roots grow from meristems found on the under-surface of each tuber. Tuberous roots, for example with *Dahlias*, produce several swollen organs growing as a unit (see Figure 6.32), with a central stem that forms shoots, leaves and flowers (see Figures 7.5 and 7.26 to 7.28). Shoots may grow from nodes on the extending stem, forming a crown-like structure around the junction between roots and stem tissues. Provided a shoot is attached to each individual swollen root, it is capable of developing into a single, separate, free-living plant.

Rhizomes are specialised stems growing horizontally just below or above soil level; examples are some irises (see Figure 6.33). Shoots develop from the upper surface and roots from the lower surface of each rhizome. Usually one upper shoot dominates the growth and forms leaves and flowers. Lateral buds can grow, however, from nodes on the rhizome surfaces (see Figures 8.23 to 8.28).

Figure 6.33 Iris *rhizome showing adventitious root system and ageing tip of the rhizome*

Rhizomes extend from the apical end where the stalk and flowers form while the older parts of the stem desiccate, senesce and eventually wither and drop away. Some buds may initiate new rhizomes growing laterally from the parent stem. Eventually, these are capable of independent life and naturally sever links with the parent plant (see Chapter 8).

A few geophyte species develop from swollen hypocotyls, this is the lower part of the stem below the seedling cotyledons. Frequently, this structure is referred to as a tuber, but it is botanically distinct because the swollen storage tissues formed from the hypocotyl. Leaves and flower shoots grow from the centre of this tuber, and roots emerge from around its periphery, as for example with cyclamen.

Geophytes are capable of the metabolic processes associated with all green plants. The scale leaves of bulbs have stomata and hence provide portals for gas exchange, taking in oxygen for respiration and carbon dioxide for photosynthesis. But until chlorophyll is manufactured in the organs of geophytes, photosynthesis will not happen.

Handling bulb orders

All of these storage organs respire producing heat. Consequently, when bulbous plants of all types are received by mail orders or as purchases from retailers, they should be unpacked, placed, if possible, in a seed tray and stored in a well-ventilated, cool, darkened store. Always retain the name and description of the bulb, and plant them as quickly as possible.

Bulb diversity

A large diversity of leaf and flower types is found in geophytes. This applies in particular for daffodils and tulips where extensive classification systems based on floral characters and divisions between botanical forms close to wild species has been developed. These systems characterise variations into single, double, semi-double and multi-flowered forms, with some producing several flowers per flower stalk (botanically known as a peduncle), while others carry only single terminal blossoms. Descriptions of these horticultural classifications are published in most reputable bulb company catalogues. The extent of variation brought about naturally or by artificial selection and breeding is immense. Some flowers develop leafy scales or scapes behind each flower, or these may be absent. Flowers may appear before the leaves or *vice versa* or both develop together; with a few bulbous plants, leaves persist all year round. Leaves may be variegated, with many showing colours other than green. Extensive colourations of both the leaves and flowers may be caused by virus infections. These effects resulted in the "Tulipmania" craze of the 17th century, especially in the

Netherlands, during which bulbs were traded for astronomical sums of money and traders' reputations made and lost.

Bulb roots

The roots of bulbs are of three types; contractile forms which pull the organs deeper into the soil helping avoid drought stress; feeder roots which provide anchorage, and root hairs which form a massive absorptive surface for water and nutrients intake. Some bulbs and corms of *Muscari*, *Tulipa* and *Crocus* lack root hairs, relying on the feeder roots for water and nutrient intake. Root numbers are determined by the quantities of meristems or primordia present on the basal plates, a larger basal plate giving rise to increasing numbers of roots. The basal plate should, for this reason, be protected from physical or chemical injury, infection by pathogens or from browsing by pests, which will diminish root formation.

Geophytes evolved worldwide but largely in those regions with a "Mediterranean" type of climate characterised by cool, wet winters and hot, dry summers. Frequently, geophytes are spring- or autumn-flowering, a rhythm evolved because it was compatible with the climate. Some, such as the autumn colchicum, flower late in the year but grow leaves in the spring, which again is an environmental response. The geographical origins vary in terms of temperature, altitude, relative humidity and soil types which, when integrated with considerable natural variations in the cold, heat and drought, resulted in the evolution of the enormous array of geophytes available for a multiplicity of uses in the garden.

6.2 Why bulbs flower

Bulbs, corms and rhizomes are large pieces of vegetative tissue allowing the parent plants capabilities for asexual reproduction and opportunities for survival. Asexual reproduction means that the progeny are replicas of the parents and known as clones (see Underpinning knowledge 3.1). This contrasts with sexually produced progenies which vary from the parents. But the characteristics of bulbous plants are also affected by the environment in which they are grown. This means that batches of bulbous plants of a particular variety will vary in their characteristics as a consequence of being propagated in areas with differing climates and soil conditions. This can cause significant differences in vigour, flower numbers and capabilities for survival between batches of bulbs. For example, *Narcissus* bulbs grown commercially in Kincardineshire, Scotland, are considered hardier, by bulb traders, compared with those grown further south, because they are exposed to harsher winter conditions even though they may be botanically the same variety. Because of this variability related to the habitat in which bulbs are grown, they should be bought from reputable and well-established sources.

Bulb appearance and quality are most important. Batches of daffodils, for instance, should contain a good proportion of double bulbs (double-nosed) (see Figure 6.12), feel firm, especially at the shoulders, with solid healthy leaf scales and without the appearance of roots from the basal plate. Signs of rooting indicate that bulbs have been stored in over-warm conditions and growth has started, and in some circumstances, this may mean that the flower shoot has aborted. Bulbs that feel soft are frequently infected with fungal rots and will either not grow or simply produce a few malformed leaves and no flowers. If the bulbs are being bought from a garden centre, ask for one bulb to be sectioned longitudinally, the scales should bright and clean without stains and a flower should stand prominently in the centre. Tulip bulbs should always be encased in a tunic which is bright and shiny. If the tunics have split and fallen away, then probably the bulbs have been kept in warm conditions.

Growing periods of bulbs

Bulbs destined for the garden should be planted as quickly as possible after purchase. The growing period for most bulbous plants is from the late winter though springtime and into the early summer. Although the roots will commence activity several months prior to late winter. After their active growth phases, the leaves senesce and the plants enter a period of dormancy. Especially with early-flowering and maturing bulbs the growth rhythm in a garden, is biologically inefficient because they enter senescence before conditions in cool temperate climates encourage maximum photosynthesis. Consequently, especially with tulips the bulbs are frequently exhausted of resources by the demands of flowering sinks and there is not sufficient time for resource replenishment into the bulb.

Consequently, the gardener should consider that bulbs used for display purposes are suited for one season of use only. Replenishing resources means that the exhausted bulbs must be grown for several years in a nursery plot, and this takes up considerable land areas. Gardeners can be successful in using these bulbs for naturalisation but the proportion which will survive is relatively low. Where bulbs are naturalised especially daffodils, the foliage must be retained for at least 8 to 10 weeks after flowering in order that resources may be produced and stored in the bulbs. Attempts at carrying bulbs from one season into the next will be further frustrated by infections from pests and diseases, particularly where these are also vectors for viruses. An accumulation of virus infections causes reduced flowering and degeneration in vigour.

Flower initiation

The mechanisms controlling flower initiation vary with different bulbous types. Much research into bulb growth, reproduction and flowering comes from the Netherlands. Early studies classified flower initiation into five groups:

Figure 6.34 *Flower formation inside a daffodil bulb*

1. Initiation during spring or early summer of the year preceding that in which they reach flower opening and before the bulbs are lifted commercially, for example *Narcissus*, *Galanthus*, *Leucojum* (see Figure 6.34).
2. Development during previous growing periods so that flowers are present by replanting time in the autumn, for example, *Tulipa*, *Hyacinthus*, *Iris reticulata*.
3. Occurring after replanting, at the low temperatures of winter or early spring, for example, bulbous species of *Iris*.

4. Extended development taking more than a year before flowering, for example, *Nerine*.
5. Alternating with leaf formation through the whole growing period, for example, *Hippeastrum* (*Amaryllis*).

Gardeners are offered "prepared" bulbs which are capable of accelerated flowering out of season. This is particularly the case with hyacinths, which are frequently grown domestically for internal decoration in winter, thereby adding some early spring brightness. Preparing bulbs simply means that they have been exposed to temperature regimes after lifting which results in accelerated flower development. Thereafter, the bulbs have been chilled in order to "hold" the flower until the gardener places it in suitable environments. With hyacinths, this means planting the bulb in a suitable container in September for flowering at Christmas. This can be a clay (terracotta) bowl, with a hole in the base, filled with compost, or a hyacinth vase which is filled with water to a level just below the bulb. Fragments of charcoal should be placed in the water which helps prevent accumulations of algae and moulds. These containers should be placed in a cool, unlit environment for about 10 weeks, during which time the leaves will have begun emerging and the flower spike will be prominently displayed well above the foliage. At this point, the pot is brought into a warm, well-lit environment and the flower spike emerges ahead of the leaves. Bringing the plant out from its cool, darkened environment at the correct moment so that the flowers expand faster than the leaves requires expertise accumulated over several years. Preparing bulbs of all types and using these for early-season crops grown in glasshouses ahead of natural field season blooms was a substantial part of the East Anglian horticultural industry. Now the flowers are frequently grown in countries where there is ample natural light and warmth, and blooms can be produced more easily and then flown to the European markets.

Naturalising

Naturalising bulbs means planting them into lawns or rough grass where they will emerge and flower against a green grass background. After flowering, naturalised bulbs will continue growing for at least 8 to 10 weeks, replacing reserves into the bulbs. This means that the grass cannot be mown until after the bulb leaves have senesced. Consequently, this practice is really only feasible in gardens with large expanses of grass which can be left un-mown well into early summer. Displays of naturalised bulbs can, however, be dramatic. Small-flowered bulbs, especially *Narcissus* species or those varieties producing heavily scented flowers such as the 'Paper White' or 'Actea' daffodils, are well suited for naturalising. Some species of tulips can be naturalised satisfactorily. But, as discussed earlier in this chapter, using tulips in this way is more difficult. Small-flowered fritillaries will naturalise well, but establishment requires several years' growth before an acceptable display is achieved. Naturalising snowdrops or corms such as crocus or anemones is most successful when plants purchased "in-the-green" are used.

The incidence of pests and diseases may mean that naturalised bulbs and corms require replacement as the years pass. Smaller bulbs appear more resistant against these problems compared with the large-flowered hybrid daffodils and tulips. Tulip fire (*Botrytis tulipae*) is one of the biggest risks, destroying leaves by causing large water-soaked lesions (see Figure 6.16). As soon as lesions become apparent, all infected foliage should be removed and destroyed. Bulbs may be invaded by nematodes, small worm-like animals, which accumulate over time, reducing vigour and viability. Slugs and snails will severely damage bulb leaves and foliage, especially in wet weather. Biological controls are available from garden centres, mail-order catalogues or via websites and provide some protection but not complete solutions.

Bulbs require little specialist nutrition. Applications of general "GrowMore" fertiliser (30 g per square metre) should be made in the early autumn ahead of bulb foliage emergence. This is because root growth starts well ahead of leaf and shoot emergence.

6.3 What makes bulbs grow?

Bulbs, corms and tubers are complex biological structures interacting with the surrounding soil and aerial environments. Similar interactions also apply for vegetable bulbs, onions and garlic and potato tubers (see Chapter 2). Their life cycles have two phases:

- Firstly, aerial shoot growth and flower formation which result in sexually produced seed. Photosynthesis in the leaves produces new sources of carbohydrates, proteins and fats, fuelling growth, and reserve resources are stored for future seasons.
- Secondly, the growth of asexually produced progeny bulbs, corms and tubers. These benefit from photosynthates stored in the newly developing progeny bulbs, corms and tubers which grow in subsequent seasons. Bulbs that are the products of this form of reproduction may be dormant when they arrive for use in the garden. But they are living tissue, and their viability can be adversely affected by storage at elevated temperatures and excessive humidity.

Removing the scale leaves which surround corms, for example, reveals a large central mass of undifferentiated tissue which has hair-line scars girdling it (see Figure 6.30). Buds can be found on what are effectively the nodes of the stem with around the base scars from which roots will emerge as the growing season proceeds (see Figure 6.31). Once root activity starts, buds on the stem open and elongate into leaves with a central flower bud. Simultaneously, the basal region of each shoot, where it is attached onto the parent corm, starts swelling. Concomitantly, the parent corm, which is below the new one, shrinks as its nutrient reserves are depleted and translocated into the new developing shoots. Corms lifted at the end of a growing season show effects of these processes. A newly formed corm has grown on top of the old parental corm (see Figure 6.29). Multiple small corms may also form around the periphery of each major new one. Each new corm acts as a sink drawing photosynthates from the leaves. Corms may reproduce several-fold in one season but the smaller corms take several years before reaching flowering size. Where, for example, crocus corms are naturalised, progeny will develop over several years by asexual multiplication, slowly increasing the mass of early spring blooms.

Bulbs (see Underpinning knowledge 6.1) have different structures compared with corms. In bulbs, the leaf scales are modified and store photosynthate resources. The simplest bulb structure is found in tulips. The bulbs are ovoid to conical in shape, rounded at the base, which is flattened and the whole coated by a tough brown membrane. There may be the remnants of a flower stem at the top of the bulb and of roots around the basal plate. Removing the brown coat reveals the white outer leaves of the bulb inside which are numerous leaf scales growing from the basal plate. Within these fleshy scales is a bud at the base which will grow into a new bulb in favourable conditions. In the centre of the bulb are the flower bud and stalk. There may be smaller leaves furled round the flower. Bulbs that have not reached flowering size have a single central leaf which emerges in the spring.

Bulb emergence

Once flowering-sized bulbs are planted, roots start emerging. This happens from late autumn and winter preceding spring-time flowering. Bulbs start growing well ahead of any signs of bud emergence through the soil surface. Consequently, areas planted with bulbs should be cultivated with great care because the newly emerging roots are easily injured, which diminishes the quality of the resultant plant and its flower. As the flower spike emerges through the soil, green leaves unfurl, the structure extends and the plant is self-contained and capable of photosynthesis, producing new resources.

Small progeny bulbs attached to the parent swell as they use some of these new resources. But the energy requirements needed for flowering and the subsequent formation of a seed capsule depletes the reservoir of resources, and this substantially reduces the vigour of the parent bulb. Parent tulip and the progeny bulbs will not reach flowering size again for some years, during which time resources are rebuilt. Gardeners are well advised against retaining "spent" tulip bulbs from one season to the next because it is very unlikely that they will produce any flowers for several years.

Narcissus bulbs are available for gardeners in a range of sizes. These include small "offsets", progeny bulbs not yet of flowering size; "rounds", which may also be too small for flowering; "single-nosed", these are bulbs with a large robust protuberance at their tops, indicating that they are of flowering size; and "double-nosed", large bulbs of flowering size probably capable of forming two flowers, one from each "nose". Daffodil bulbs are globular or flask-shaped, having a narrow neck and flattened base; the whole is covered in dry membranous scales protecting inner fleshy scales, some of which will emerge above ground as leaves. Each scale forms from a disc-shaped basal plate. The concentrically arranged fleshy scales carry buds at their bases and may be sliced into small segments, as with tulips and used for artificial propagation in processes known as "twin-scaling" or "chipping".

Roots emerge from around the basal plate as the bulb commences growth; this happens well before leaf emergence. The compact structure of *Narcissus* bulbs means that the leaf scales grow tightly together with the flower spike emerging from the centre. As leaves extend above ground, the bulb is regenerating; consequently, parent daffodil bulbs, unlike tulips, may remain of flowering size over several seasons. Alternatively, daffodil bulbs may divide following flowering. This "splitting" means that the progeny will be too small for flower formation in the succeeding season. As with tulips, gardeners with small gardens should not take up valuable space by growing daffodil bulbs into their second and subsequent seasons. They will achieve far more satisfactory results by renewing and replacing bulbs annually which provides colourful and enjoyable gardens.

Where considerable garden space permits, daffodils, in particular, will establish and provide many years of colourful displays. Some varieties are "shy flowerers", meaning that they naturally only produce limited numbers of blooms relative to the numbers of bulbs planted. For normal garden purposes, these varieties should be avoided, and local knowledge can be very helpful in identifying these types.

Initially, flowers are the primary sinks for the products of photosynthesis (see Underpinning knowledge 4.2 and 5.2). After flowering, the leaf bases and scales in the bulb become the sinks. Once the plants senesce, the leaves and flower stalks separate from the bulb, which persists as the storage organ. The bulbs may now be larger than those originally planted in the ground. Buds will develop in the axils of leaf bases within the bulb. The bulb swells and divides into two daughters because the parent begun "splitting". "Splitting" is a horticultural term that describes the manner by which parent bulbs naturally divide into two or more progeny bulbs. The two daughter bulbs may be joined together initially by wrapper leaves. Sometimes uneven development between these two daughter bulbs results in one reaching flowering size before the other. These variations are characteristics of different daffodil varieties. For this reason, some varieties are favoured commercially compared with others. Some varieties do not separate completely, forming clumps of bulbs. This is a characteristic of species and wild-type bulbs. Individual bulbs within a clump will be flat-sided, caused by pressure from companions.

Dividing established bulb colonies

Over time, clumping often results in diminishing numbers of flowers because all of the plants are competing for resources. Consequently, clumps of bulbs or other bulbous-like types, as for example *Agapanthus africanus*, should be lifted and divided every 4 or 5 years, rejuvenating the flowering capabilities (see Figure 6.35).

Lifting and dividing also provides an opportunity for removing perennial weeds such as couch grass (*A. repens*) or ground elder (*A. podagraria*) (see Figure 4.62) where their rhizomes are deeply entwined with the bulbs. In this process, clumps should be separated into individual bulbs or other clumps (see Figure 6.36) as with other bulbous-like plants and washed

Figure 6.35 Agapanthus *clump ready for division*

Figure 6.36 Agapanthus *clump divided into three parts*

Figure 6.37 *Planting* Agapanthus *parts in a hole with compost and water*

Figure 6.38 Agapanthus *part topped up with soil*

Figure 6.39 Agapanthus *replanted and mulched with gravel chippings*

thoroughly, removing the last traces of weed rhizomes before replanting (see Figure 6.37). Once placed in the hole, the *Agapanthus* plants should have soil drawn round the neck of the plant (see Figure 6.38). Once replanted, a gravel mulch can be placed around the plant which deters weed growth (see Figure 6.39).

Pests and diseases

Infections of bulbs and corms with virus diseases result from aphid or nematode invasions. Some soil-borne fungi also invade, resulting in rotting, especially from the basal plate and towards the centre. Once such infections are established, they will spread into the surrounding bulbs. The only solution is removing all of the infected bulbs and resting the ground from bulbs for 3 or 4 years. During this time, naturally beneficial microbial processes will degrade these pathogens. This results from the actions of soil bacteria and fungi which are predators of plant pathogens. Air-borne fungi can infect bulbs and are seen as watery patches on the bulb or on leaves and flowers. Spraying with proprietary fungicides can provide control and applications should be repeated during the growing and flowering season, following the manufacturers' instructions. The most reliable insurance against these problems results from purchasing bulbs and corms only from reliable and well-established suppliers, avoiding traders at shows and flower markets.

Achievements: have you understood and learnt?

At the end of the chapter, you should be able to:

1. Understand that bulbs, corms and rhizomes, referred to collectively as "bulbous plants", are purchased from garden centres, mail-order companies or specialist nurserymen, and that bulbs and corms, in particular, should be accompanied by a guarantee of being "virus-tested".
2. Know that on receipt, these plants should be removed from their packaging, and if they cannot be planted immediately, they should be stored in a cool, darkened, vermin-proof place.
3. Know how to prepare land for bulb planting, usually in late summer or early autumn, and that a planting plan should be prepared in advance.
4. Know that each bulbous plant should be treated carefully, avoiding damage and placed into an individual hole in the soil.x
5. Know that bulbous plants, particularly daffodils, tulips and hyacinths, may be planted into containers for decorating areas of hard landscaping or for internal decorations, and the processes required in order to achieve this.
6. Know that bulbous plants will begin emerging in borders and from outdoor containers in the early spring and that these will require support and that containers should be watered at regular intervals even when there have been periods of rain.
7. Understand that after flowering the residual flower heads should be removed in order that the parent bulb may continue growing and that the foliage should be retained for at least 8 to 10 weeks before being removed.
8. Know that borders where bulbs remain for several seasons may be carefully planted with annual bedding plants, alternatively that bulbous plants may be naturalised into grassed areas.
9. Understand that some bulbous plants such as crocus, anemones and aconites are most easily established from supplies obtained "in-the-green" as growing plants and the processes required for their establishment.
10. Understand that bulbous plants such as those growing from rhizomes may also be purchased as growing plants from garden centres or specialist nurserymen and the treatment these should receive before planting out, during planting and the after-care required.
11. Describe the parts of bulbous plants such as bulbs, corms and rhizomes and the diversity of foliage and flowers associated with bulbous plants.
12. Describe the source-sink relationships existing between foliage, flowers and bulbs, corms and rhizomes.
13. Understand how some bulbs, especially hyacinths, may be "prepared" by specialists for later use, accelerating flowering for early indoor decorations.

14. Understand that, since bulbous plants are large pieces of vegetative tissue capable of surviving for several seasons, they may accumulate pests and pathogens which damage the prospects for future growth and flowering.
15. Describe the phases of growth, flowering and senescence of different bulbous plants and how these are used advantageously by the gardener.
16. Describe how bulbous plants reproduce asexually and the importance of this process for gardeners.

Definition of terms: *Describe* is able to show why a scientific principle is important; *Understand* applies scientific principles into the context of gardening; *Know* carries through scientific principles in practical gardening activities.

Chapter 7
Flowering plants

Introduction

Herbaceous flowering plants provide much-valued colour, form, structure and enjoyment in the garden. The traditional long herbaceous border is one of English gardening's greatest triumphs. It was perfected through the 19th and into the early 20th century when, with great exuberance, larger gardens cultured an astonishing array of plants of varying heights, providing an eye-catching summer-long display. This required a great deal of knowledge, work and considerable ingenuity. Consequently, the fashion waned as the supply of skilful gardeners was diminished by two world wars and subsequent economic and social changes. Today, hobby gardeners of modest means replicate herbaceous borders, often achieving stupendous displays, and continue enhancing urban and rural areas by their ability and determination. Some of these gardens are opened for public enjoyment as part of the National Gardens Scheme (NGS). This organisation advertises local opening times based on a county system and the modest charges benefit a range of local and national charities.

Garden practices

a. Types of herbaceous plant

Herbaceous plants are challenging because after their floriferous displays they senesce in the autumn, and once dead foliage and flowers are removed, the ground is apparently bare and uninteresting. There are three forms of herbaceous plants as defined by their growth in cool temperate climates:

1. Annuals which grow and flower for a single season, and these may be:
    a. Hardy annuals, which resist low temperatures and can be relied upon for an early display and may be planted or sown *in situ* as soon as soil temperatures rise above about 5 °C, permitting root growth in the early spring (see Figure 7.1).

Figure 7.1 *Cornflower* (Centaurea cyanus), *hardy annual*

Figure 7.2 *French marigold* (Tagetes erecta x patula), *half-hardy annual*

These native wild flowers have been selectively bred recently, with the result that the colour range has been increased from the traditional "wild flower" blue and with the addition of some dwarf hybrids. Some perennial forms, for example 'Jody', are indicative of the blurring of some of these categories as a result of gardening in varying localities and seasons.

b. Half-hardy annuals may have some resistance to low temperatures but are killed by prolonged periods of frost and need protection if planted early in the season (see Figure 7.2).

c. Tender annuals will provide an exuberantly colourful display but are killed by even the slightest frost or even damaged by chilling when temperatures drop below 5 °C. Consequently, they require considerable protection if planted before the danger of frosts has passed (see Figure 7.3).

Plants which require treatment as annuals in northern climates may originate from warmer temperate or tropical zones where they grow as huge and floriferous perennials.

2. Biennials are herbaceous plants which develop vegetatively from seed in the first season after they have been sown, forming a robust structure surviving through the following winter and flowering in the next season. Low temperatures vernalise these plants, stimulating their flowering (see Underpinning knowledge 5.2 and 5.4) (see Figure 7.4).
3. Perennials are herbaceous plants which, once established, will re-emerge each spring, re-growing from a dormant storage organ, and provide continuing pleasure for the gardener. Frost-tender perennials require storage over winter (see Figure 7.5) in protected conditions.

Perennials require similar husbandry and nutritional care as provided for soft fruits (see Chapter 5). Some biennials and perennials may be treated as annuals even though they evolved as longer-lived plants in their original environment. This demonstrates the flexibility of ornamental plants for a multitude of uses steered by skilful gardening.

Figure 7.3 *Sunflower* (Helianthus annuus), *tender annual*

Figure 7.4 *Wallflower* (Cherianthus cheiri), *hardy biennial, note the "bee-friendly" label indicating their helpfulness for pollinators*

Figure 7.5 Dahlia *variety 'Bishop of Llandaff', a tender perennial*

b. Maintaining herbaceous perennials

Careful removal of weeds is particularly important since these build up as hardy herbaceous perennials mature and expand. Consequently, every few years it is necessary to lift these plants and divide the clumps, teasing out weed roots and re-planting the now smaller bundles. Since the volume of stock of herbaceous perennials increases each time the clumps are lifted, they are exceptionally good value for money.

A good example of this process is the division of herbaceous *Campanula* plants. Lift the entire plant with a digging fork with a periphery of soil about 5 cm beyond the diameter of the clump (see Figure 7.6). These can be sizeable, especially if they have been in position for some years and hence quite heavy. Place the clump on a mat of heavy gauge plastic sheeting or an old hessian sack, preventing damage to the lawn. Drive two digging forks placed back to back down through the centre of the clump and force their handles together (see Figure 7.7).

Figure 7.6 *Lifting a clump of* Campanula *spp. for splitting*

Figure 7.7 *Splitting a clump of* Campanula *with two forks placed back to back*

That action leavers the tines apart, splitting the clump. Repeat this process as often as required until the original clump is reduced into a series of small pieces. Examine each piece carefully identifying *Campanula* roots and those of any unwanted fragments of weeds (see Figure 4.62, for example, ground elder, *Aegopodium podagraria*). Carefully disentangle the weed and *Campanula* roots, making sure that no unwanted growth is left. Ensure that the roots of *Campanula* plants do not dry out by placing all the new plants into a bucket of water. Weed roots should be removed and destroyed.

The new *Campanula* plants are replaced into the border in arrangements that are compatible with the planting plans. Make suitably sized holes with a spade or trowel, place some general garden multipurpose compost in the bottom and mix this with the soil with a digging or hand fork. Pour water into the hole, ensuring that this drains easily. Place each *Campanula* plant into a hole such that its roots are resting on the compost but the neck of the plant where the roots and leaves meet is level with the soil surface; refill the hole with friable soil and firm round the plant (see Figure 7.8).

Lightly fork over the soil around the plants, removing footprints and adding more soil and compost as required. Sprinkle general-purpose "GrowMore" fertiliser around the plants and water with a watering can fitted with a coarse rose (see Figure 7.9).

Place slug bait around the plants, as the new young succulent growth will be an attractive food source. Where there are concerns about using slug bait pellets, these may be contained under a "bell" which prevents access by birds, pets and children. Monitor the plants at regular intervals, keeping them watered, especially during periods of dry weather. Examine the plants from time to time, ensuring that they have not been disturbed; cats looking for soft earth can

Figure 7.8 *Planting a sub-division of* Campanula *in the hole with compost and fertiliser*

Figure 7.9 *Campanula sub-division planted and watered well in the border*

be a nuisance. The new plants will recover from division quite quickly and commence re-growing. During the following spring, growth will progress and they should flower by mid-June. In all probability, the display of flowers will be more vigorous compared with that produced before division took place. This is a benefit obtained from re-propagation and the stimulation of new growth.

c. Planning herbaceous borders

Planning herbaceous borders requires considerable forethought and understanding. Primarily, plans should establish a colour scheme which suits individual preferences and with mixtures of plants that complement rather than compete in the border. The art of ornamental gardening lies in achieving compatibility of plant height, leaf form, architectural structure, seasonality and, most importantly, flower colours with their lengths of display. There is no point in designing borders where plants flower out of synchrony and are either so vigorous that they overgrow all the other plants or alternatively are so slow-growing that they are lost by competition coming from other plants.

Gaining appreciation of plant compatibilities or incongruities comes from accumulated knowledge, experience and skills; it is achieved over a considerable period of time. Apparent mistakes are rectified by moving plants which have been placed wrongly or by adding newer ones that extend an effect. Gardening is far from a static art form. It is one with considerable opportunities for making changes and re-organising the canvass using live material. Gardeners gain much by visiting borders which are already in place and functioning, such as National Trust gardens or the great borders at the Royal Botanic Garden Kew or the RHS Wisley Garden. Ideas may also be gleaned from designs at some of the RHS shows such as the Chelsea Flower

Show or from more localised displays. Remember, however, that designers in these competitions frequently use plants solely for an instant effect, and such combinations lack practical reality because they are used out of their normal seasons.

Increasingly, there is a huge range of herbaceous plants available for gardeners. This reflects their popularity and the ease with which many are propagated commercially. Plant breeders are providing new colourful plants which will withstand adverse weather conditions and neglect each year.

Plants at various stages of growth are available from garden centres or from mail-order companies. Some arrive as very young material requiring further husbandry before being transplanted into the garden (see Figures 7.10 and 7.11).

Others may be purchased as mature plants and are ready for immediate establishment in the border (see Figures 7.12, 7.13 and 7.14).

Figure 7.10 Geranium *seedlings in modular trays in a garden centre*

Figure 7.11 Lobelia *seedlings in modular trays in a garden centre*

Figure 7.12 *Variously coloured primrose plants in pots in a garden centre*

Figure 7.13 *Packaged cowslip plants in pots and sent by mail order*

Over-eagerness in purchasing early in the season should be avoided because the plants will have been raised under protection and in consequence will suffer frost or cold damage if planted too soon. Potted herbaceous plants can be used in attractive plantings in tubs and urns for patios and piazzas, for example the use of zonal *Pelargonium* plants (see Figure 7.14).

Planting procedures are similar to those described in Chapter 6 for bulbs. Containers should provide ample drainage. If half-barrels are purchased, then holes should be drilled in the base, if this has not been done by the vendor (see Figure 7.15).

The root ball of each plant should be severed with a sharp gardener's knife (see Figure 7.16) which encourages rapid growth into the compost into which they will be planted.

Arrange the plants in an attractive pattern in the container, planting closely together so that a vigorous display of flowers is achieved quickly (see Figure 7.17). *Pelargoniums* will provide colour and interest on patios for the summer (see Figure 7.18). They will require regular watering in the period after planting. Thereafter, watering should be restricted and they should not be provided with added fertiliser. Stressing the plants encourages flowering. Nonetheless, they should not dry out to the point of wilting, senescent leaves and flowers should be removed as these become infected with grey mould disease (*Botrytis cinerae*) which is unsightly and shortens the period of display. A rotation of *Pelargonium* and spring bulbs, especially tulips such as varieties of the 'Apeldoorn' type, provides a colourful display from early season to late autumn frosts.

There is often confusion between *Geranium* (Cranesbill) and *Pelargonium* (Storksbill). The former originate from parts of Asia, Europe and northern Africa, while the latter come mostly from South Africa. Botanically, they are closely related but can be separated only by slight differences in flower structure

Figure 7.14 *Zonal* Pelargonium *in a pot ready for planting out*

Figure 7.15 *Holes drilled in the bottom of a half-barrel, increasing drainage in preparation for planting with tender display plants*

Figure 7.16 *Severing the root ball of Zonal* Pelargonium *in preparation for planting*

Figure 7.17 *Planting a Zonal* Pelargonium *plant in a half-barrel*

Figure 7.18 *Half-barrel filled with Zonal* Pelargonium *plants for a summer display*

Figure 7.19 *Perennial* Geranium *spp. growing in a border*

and habit. In the garden, *Pelargonium* tend to be frost-susceptible, while *Geranium* are hardy, forming large, colourful clumps in the border (see Figure 7.19). Herbaceous plants such as perennial *Geranium* spp. may succeed bulbs in a border providing a continuity of display almost the year round.

d. Growing herbaceous plants

The garden pansy (*Viola* spp.) is a good example of a popular hardy herbaceous plant. Plant breeders have greatly improved garden pansies over the last decade or so. There are now forms that can be grown almost year-round. Flower size varies from the very large to minutely small types available in a range of blossom colours. Growing pansies from seed is very similar to those processes already described for small-seeded vegetables (see Chapter 4). Select the type of pansy which is most attractive and fits in with your border design plans. Alternatively, they may be purchased from garden centres or by mail order.

The simplest manner for germinating pansy seed is using modular trays, as previously described for small-seeded vegetables (Chapter 4). Open the packet and probably the inner foil container and place the seed in a small open plastic saucer and follow the procedures of Chapter 4.

Monitor the growth of the seedlings closely. As the plants develop, several leaves remove a module and examine the root growth. Push the module from its container by gently inserting a piece of bamboo cane into the hole in the base of the tray. The roots should be healthy and white growing in sufficient numbers that they fill the module but without giving an overcrowded appearance. If the leaves show signs of yellowing (chlorosis), additional nutrients are needed. Prepare a very weak, well-diluted solution of tomato feed (for example 1 part tomato feed to 10 parts water) which has a high potash value and minimal nitrogen formula on the package. It is essential that the nutrient solution is well diluted, as seedlings will be scorched by full-strength nutrient. Ensure that leaves are washed over, removing any traces of nutrient solution so that damage does not develop.

Once the seedlings reach a sufficiently large size, possibly showing some flower buds, they are transplanted into the display border. Previously, the border should have been cultivated and prepared in a manner similar to that used for the vegetable crops (Chapters 3 and 4). Herbaceous plants should be placed out in designs, not in the straight rows used for vegetables. This involves transferring previously prepared planting designs onto the ground, tracing them out with bamboo canes and fine sand as markers.

For each plant, take out a hole with a trowel which is sufficiently large so that it accommodates the roots of each pansy transplant, as described in Chapter 4. Once all the plants are in place, water well and add a top dressing of "GrowMore" general fertiliser at 30 g per square metre. Ensure that any fertiliser granules that lodge on leaves are washed off with a watering can fitted with a coarse rose. Treat the ground with slug repellent and place a network of black cotton attached to bamboo canes above the plants in order to deter pigeons and other birds which will find the leaves are an attractive food source (see Figure 7.20).

Figure 7.20 *Pigeon damage in a border filled with annual bedding plants*

Alternatively, place plastic netting over the border to keep out birds; pigeons in particular are menaces, causing freshly planted herbaceous plants considerable damage. In some areas, deer and rabbits may be attracted by herbaceous plants, requiring a barrier of wire netting. Keep the plants well watered and remove dead flower heads and any yellowing leaves. This lengthens the display and deters diseases. It may be necessary to apply sprays occasionally, deterring pests and diseases. Pansies are best considered as useful for only one season and when growth and flowering begin waning they should be replaced. They make good background planting for bulbs such as tulips and, in this setting, should be removed with the bulbs.

Raising annuals from seed is a time- and space-consuming exercise; consequently, buying module-raised plug plants from a garden centre or from a mail-order supplier is an attractive, labour- and time-saving alternative (see Figure 7.21). These plants vary in size and cost, from small seedlings to larger specimens capable of being put out directly into a planting design (see Figure 7.22).

Small plants may require a period of growing-on under protection. In that case, transfer each module into a 6 cm diameter plastic pot. Fill the pot half full of potting compost and hold the plug plant carefully in the pot with the left hand, and with the right hand sprinkle compost around the plug. Fill the pot to within 1 to 2 cm of its rim. Gently firm the compost around the neck of the plug plant, carefully avoiding the stem. Place the pots on greenhouse benching in a pot-thick arrangement. That means each pot should be touching others within the row and between the rows. That economises on bench space, simplifies watering and encourages robust growth. Growing plants "pot-thick" in the early stages encourages uniformity, but care is required so that the plants are not weakened by competition for light as they become larger.

Figure 7.21 Viola *variety 'Blackberry' plants in a modular tray from a garden centre, requiring growing-on before planting in the garden*

Figure 7.22 Viola *variety 'Blackberry' modular plant as bought from a garden centre, showing the shoot and root systems*

Spray the pots with water from a can fitted with a fine rose, completely soaking the compost. The level of compost may settle in the pots, in which case add a small amount more around the necks of the plants and give more water.

Monitor these plants regularly, preventing the compost from drying out. Avoid over-watering as that encourages leaf-, stem- and root-rotting diseases. As the plants grow, space out the pots, encouraging air movement around the plants. When growth is such that the plants crowd each other for a second time, they will be ready for transplanting as previously described. These will provide a colourful and lengthy display in a border (see Figure 7.23).

Besides pansies, there is a range of hardy annual plants which are valuable for garden borders; half-hardy annuals are also valuable additions for borders (see Table 7.1). These resist mildly chilling temperatures but are damaged by leaf burn or flower abortion. Plants bought from garden centres or mail-order companies have been raised under protection in glasshouses and polythene tunnels. Consequently, they are very tender and susceptible to even light chilling damage. They can be hardened off by placing them in the greenhouse for a couple of weeks before transplanting. They will require careful monitoring, ensuring that they are well watered during this time. Quite possibly, the commercial propagator will have grown them on benches fitted with automatic root zone irrigation. That means the plants will be damaged by water shortages.

Judging the correct time for transplanting depends on the garden's locality since localised microclimates may vary substantially. Country gardens, for example, suffer from frost damage prolonged later into the season than those in neighbouring urban and suburban areas. Wind

Figure 7.23 Viola *variety 'Blackberry' planted out in the border*

direction may also influence frost occurrence. This is especially a problem if the garden is situated in a frost pocket at the bottom of a slope. Normally, it would be expected that half-hardy annuals may be transplanted in the southernmost counties of England at the beginning of May and with delays progressing northwards until planting takes place in the first weeks of June or even later in the far north of Scotland.

Fully tender plants make excellent "dot-plants" in borders acting as foci of interest (see Table 7.1). These cannot be transplanted until all risk of frost is over. Then they should be planted centrally in positions reserved for them as large plants bought especially for that purpose. Raising these plants requires a heated glasshouse and husbandry over a long period of growing time, which is probably best left as a job for professional horticulturists.

Table 7.1 *Herbaceous ornamentals and their approximate hardiness (this varies considerably depending on location)*

Hardy annuals	Half-hardy annuals	Hardy biennials	Hardy perennials	Tender specimen plants
African daisy (*Osteospermum*)	*Ageratum*	Canterbury bells (*Campanula medium*)	*Alyssum*	*Begonia*
Agrostemma (*Lychnis*) – various campions	*Alyssum maritimum* – can be longer lived	Daisy (*Bellis perennis*)	*Aquilegia* (columbine)	*Dahlia*
Candytuft (*Iberis*)	*Aster*	Foxglove (*Digitalis*)	*Campanula* spp.	*Gaura*
Celosia	*Antirrhinum* (snapdragon)	Forget me not (*Myosotis*)	Chinese lantern (*Physalis franchetti*)	*Geranium* spp.
Clarkia	Californian poppy (*Eschscholzia*)	Sweet rocket (*Hesperis*)	*Doronicum*	*Geum*
Calendula (Scotch or pot marigold)	*Cineraria maritima*	*Hibiscus* rose mallow)	*Echinacea*	*Mimulus*
Cornflower (*Centaurea*)	*Cosmos*	Honesty (*Lunaria*)	*Delphinium*	Castor oil plant (*Ricinus communis*)
Felicia	*Gazania*	Pansy* (*Viola*)	Hollyhock (*Alcea rosea*)	Sunflower
Gaillardia	*Gerbera*	Sweet William (*Dianthus barbatus*)	Larkspur (*Delphinium ajacis*)	
Gilia	Cherry pie (*Heliotrope*)	Wallflower	*Leucanthemum*	
Godetia	*Lobelia*		Lupin	
Gypsophila	Livingstone daisy (*Mesembryanthemum*)		*Penstemon*	

Hardy annuals	Half-hardy annuals	Hardy biennials	Hardy perennials	Tender specimen plants
Larkspur	*Madia*		Pansy* (*Viola*)	
Linaria	*Matricaria*		*Perovskia*	
Linum	Morning glory (*Ipomoea*)		Oriental poppy (*Papaver orientalis*)	
Love-lies-bleeding (*Amaranthus caudatus*)	*Mirabilis*		Polyanthus	
Love-in-the-mist (*Nigella*)	*Nemisia*		Primrose	
Nasturtium	*Petunia*		Pyrethrum (*Chrysanthemum coccineum*)	
Oxalis	*Phlox drummondi*		*Rudbeckia*	
Phacelia	*Portulaca*		*Scabious*	
Field poppy (*Rhoeas*)	*Salvia*		*Verbena bonariensis*	
Sweet pea	Butterfly plant (*Schizanthus*)		*Aubretia*	
Virginian stock (*Malcolmia maritima*)	*Statice*		*Armeria*	
	Stocks		*Agastache*	
	Tagetes			
	Torenia			
	Verbena (Vervain)			

Notes: Listed according to most common use as described by several seedsmen in their catalogues; *pansy may be used as a biennial or perennial

e. Biennial herbaceous plants

Biennial plants add greatly to the colourfulness of borders and can be especially useful in the early spring. Wallflowers (*C. cheiri*) are one of the most popular biennials with a huge range of colours and substantial scent. They are very suitable for mixed planting with bulbs, especially tulips and daffodils. Raising wallflowers from seed requires good planning. Wallflower seed can be germinated and initially propagated in the glasshouse using the techniques outlined for small vegetable seed (Chapter 4). They belong to the brassica group of plants and can be treated as described for cabbage in Chapter 4. Alternatively, seed can be sown directly into the garden, again using the techniques outlined for vegetables (Chapter 4).

The ideal place for sowing wallflower seed and producing plants is the vegetable garden. The seed should be sown in May when much of the early-season work in the vegetable garden has been completed and crops are growing. The wallflower plants will develop through the summer in preparation for their later use in the flower garden. The plants will become quite vigorous so that it is advisable for them to be spaced at about 7 to 8 cm within the row and with about 10 cm between the rows. Like all brassicas, they require ample nutrition so that applications of "GrowMore" general fertiliser at sowing should be supplemented in mid-July with a top-dressing of a quick-acting fertiliser such as calcium nitrate or using a liquid feed. There are numerous proprietary liquid-feed formulations available. Seaweed-based products are ideal for this purpose. These deliver a range of nutrients and other compounds which encourage robust, healthy growth, providing a basis for vigorous flowering next spring.

Wallflower plants should be lifted from the rows in the vegetable garden in mid-September for planting out in a flowering scheme mixed with bulbs. Placing the tines of a fork under the plants and gently easing them from the soil, ensuring that as many roots as possible are retained (see Figure 7.24).

Those purchased in garden centres may have less robust root systems because they have been "pulled" from the propagation bed. In either case, place the roots in a bucket of water. Plant as described for other crops and bulbs using a trowel (Chapters 4, 5 and 6). Transplanting should take place as quickly as possible; if there is a delay, stand the bucket containing the plants and water in a cool shed or garage. Once they are planted, water well using a can fitted with a coarse rose.

Wallflowers are quite hardy and will make substantial growth once warm spring weather returns; consequently, at transplanting there should be spaces of 15 cm between plants. Bulbs can be planted quite densely between them. Sprinkle some slug bait around the plants. The border containing bulbs and wallflowers should be monitored carefully and weed seedlings removed. Ensure that the border soil does not dry out causing the plants and bulbs drought stress. The bulbs will not appear until late January or early February. Wallflowers require a period of low temperatures, which initiates the flowering buds (vernalisation) and will benefit from winter frosts (see Underpinning knowledge 3.2 and 5.2). They make very colourful displays and are frequently used for reliable street planting (see Figure 7.25). There is now a huge range of

Figure 7.24 *Bare-rooted wallflower* (C. cheiri) *plants for direct planting in the border*

Figure 7.25 *Wallflower* (C. cheiri) *plants well established in a border and flowering (mixed colours)*

attractive biennials for use accompanying spring bulbs or used as displays in their own right, as listed in Table 7.1.

f. Perennial herbaceous plants

Several years of cultivation are required before herbaceous perennials are sufficiently large for use in borders. It is feasible for gardeners to grow their own plants from seed or from cuttings, but this is a long-term task requiring that a section of the garden is set aside for use as a nursery bed. Consequently, gardeners may prefer buying selected perennials from a garden centre or by mail order from specialist producers. Avoid buying "special offer" plants at discounted prices, as these will probably be small plants which demand considerable care before they can be planted into a border. In gardening, the adage "you get what you pay for" is very applicable.

High-quality plants will be sold in containers in which nurserymen have cultivated them. Inspect the plants carefully, ensuring that they are not showing signs of distress caused by erratic watering, diseases, pest damage or nutrient deficiencies. Symptoms of erratic cultivation include: browning of the foliage, rotting of leaves or at the base of the plant, yellowing and discolouration, flower abortion and a generally unthrifty appearance. Useable plants will show vigorous growth with ample flower buds. Vegetative buds and shoots on the plant should have a strong and robust appearance. Do not buy plants too early in the season as these may have been forced under glass or plastic protection in readiness for sales which coincide with particular holidays and festivals. Once planted in the garden, they will suffer badly from any adverse weather and not provide a good display. Garden centres and other retailers sell specific types of plant when they are providing the most attractive and visually eye-catching display for them. These may not be the most suitable plant for a particular garden because of its soil type, location or the wishes of the gardener.

Some herbaceous perennials can be bought as dormant tubers such as *Dahlias* (see Figure 7.26). These require potting-up in multipurpose compost for 4 to 6 weeks and grown in a greenhouse before they are planted into the garden when there is no likelihood of frosts. During that time, they will produce substantial foliage (see Figure 7.27).

In the garden, they will provide exuberant displays that continue right through until the first frosts. Displays are encouraged by removing flowers as they fade, providing support for the foliage with bamboos canes and 4-ply twine, and watering in dry spells. *Dahlias*, like many herbaceous perennials, are very attractive for slugs; consequently, they require protection with bait or bio-control. Feeding should be provided with caution. Adding nitrogen will encourage leaf and stem growth at the expense of flowering. Use liquid feeds that have high potassium and

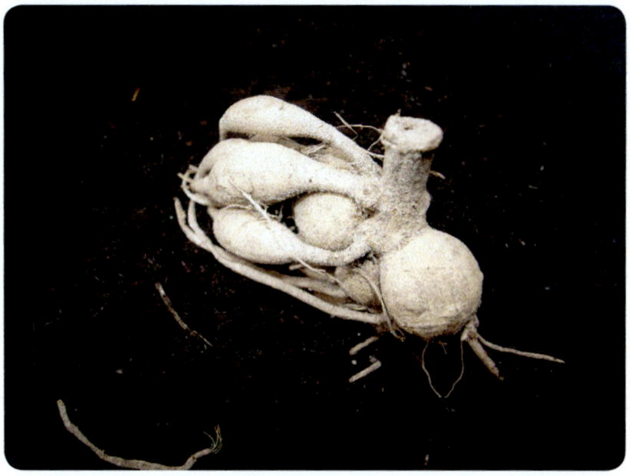

Figure 7.26 *Overwintered* Dahlia *tubers ready for potting-up*

Figure 7.27 *Well-developed* Dahlia *plants ready for planting out in a border*

Figure 7.28 Dahlia coccinea *var. palmeri in flower, this is one of the few semi-hardy perennial dahlias for the border*

phosphorus contents and minimal nitrogen ratios. In autumn, the tubers should be lifted, the foliage removed and allowed to dry out. Thereafter, store the rhizomes in a frost-proof shed or garage and placed well away from possible damage by rodents.

Some *Dahlias* are frost tolerant at least in southern counties for example *D. coccinea* var. palmeri which produces a superb display of orange flowers and reputedly will withstand temperatures as low as –10 °C (see Figure 7.28). Regrowth in early summer is slow; consequently, spreading slug bait or bio-control where they have been planted is a wise precaution against the emerging shoots being browsed.

g. Planting and subsequent husbandry

Buying herbaceous plants after about mid-May requires care and caution since they will suffer stress in periods of low rainfall and not make sufficient root growth which is required for successful growth and flowering. Place purchased plants into buckets of water, ensuring that the roots are given a thorough soaking. This wets the whole root ball and settles the roots after disturbance caused by being transported from the plant-raising nursery to a garden centre and finally into the garden.

Select planting places well in advance. The ground should have been dug during the winter or early spring and received liberal amounts of well-rotted compost or farmyard manure. Take out a planting hole with the spade which is large enough for whole root ball of the plant. This can be tested by standing the plant in its pot in the hole, and the rim of the pot should be at the same level as the top of the hole. Add compost into the bottom of the hole mixing it into the soil with a fork. Mix the compost and soil, ensuring there is ample aeration at the bottom of the hole. Pour about 4 litres of water into the hole and allow this to drain. Remove the plant from its pot. This is done by reversing the plant in your hand with two fingers placed either side of the neck of the plant ("neck" is the junction between aerial growth and where the root system has developed). Gently ease the plant from the pot. Tapping the pot on the side of the garden barrow will help separate the plant and pot. Plants that only reluctantly leave their pots may require more robust action such as cutting the side of the container. Cut into the root ball both along its sides and at the bottom with a sharp knife.

Healthy roots will be white, if they are brown and tangled then that suggests that the plant is "pot-bound", having been grown for too long in the same container. These roots will grow and penetrate the surrounding soil only very slowly, requiring continual attention, ensuring that they are watered often during spells of dry weather. Roots of very badly "pot-bound" plants may never fully emerge from the root ball and will fail even after several years in the garden. It is unfortunately not uncommon for a plant surviving in the garden for one or even two seasons and then failing for no obvious reason. Digging up such plants will probably show that they have not made any root growth beyond the old root ball. High-quality garden centres provide certificates guaranteeing that failed plants will be replaced.

Place the plant in the hole with the top of the root ball just below the soil surface; this resembles the techniques described in Chapter 5 for raspberry and blackcurrant plants. Add multipurpose compost and soil round the plant, firming with the fingers, lift the plant several times, agitating compost and soil in place around the roots. Finally, add soil and compost on top of the root ball and carefully but firmly tread round the neck of the plant consolidating it. This helps prevent "wind-rocking" of newly planted perennials. "Wind-rocking" occurs when a plant is buffeted by wind, and this causes the formation of a void in the soil around the roots which can fill with water, asphyxiating them.

Sprinkle more soil and compost and pour a full bucket of water, about 5 litres, round each plant. Add a top dressing of "GrowMore" general fertiliser around the plant and lightly fork this into the soil, then place slug bait around the plant. Young succulent leaves of herbaceous plants are very attractive for slugs. Drive three or four equally spaced stout bamboos canes around the plant. Use canes which are long enough so that they will accommodate growth by the plant in a couple of months' time. Tie round the canes with 4-ply garden string at the point where all the foliage is held inside the canes; more support will be needed as the plant grows and flowers.

When the season ends, the foliage will turn yellow. This should be removed using a strong, sharp pair of secateurs, and at the same time clear away the bamboo canes, leaving a small stake which indicates the plant's location for future years.

Early in the next spring, lightly fork over the border down to 6 to 8 cm, remove all weeds including their roots and add "GrowMore" general fertiliser at 30 g per square metre, and replenish dressings of slug bait. Monitor the border for regrowth of the herbaceous perennials, once shoots appear replace the bamboo supports into the border. Pea boughs can be used as an alternative and they eliminate the need for string ties. Regular light forking discourages weeds and increases soil aeration. Established borders can be mulched with composted wood chips, which reduces weed growth and retains moisture in the ground. Alternatively, borders that are planned in advance may be covered with porous fabrics through which planting takes place. The fabric will prevent weed growth and save considerable work.

Robust border perennials include forms of *Penstemon* (see Figure 7.29) and *Delphinium grandiflorum*. These are beautiful border plants with a wide range of forms and increasingly an extended colour range. Some dwarf types of *Delphinium* are very suitable for smaller gardens (see Figure 7.30).

The Japanese anemone (*Anemone hupehensis*) (see Figure 7.31) is a very reliable and hardy perennial suitable for most gardens and soil types. They grow very vigorously, producing

Figure 7.29 *Well-developed* Penstemon *var. 'Pentastic Red' suitable for border planting*

Figure 7.30 *Dwarf* Delphinium grandiflorum *var. 'Blauer Zweig' suitable for border planting*

Figure 7.31 *Japanese anemone* (Anemone hupehensis)*, hardy border perennial*

Figure 7.32 *White phlox* (Phlox paniculata)*, hardy border perennial*

masses of pink or white flowers, with some double forms currentlynow on the market. Phlox is similarly suitable for herbaceous borders as a hardy plant (see Figure 7.32). The colour range extends through white, pink and blue flowers which are produced in profusion towards late summer with a fragrance which is a delight on warm evenings.

Unlike the florists' sweet pea, which are grown annually and require considerable attention in terms of nutrition and support structures, the everlasting pea, *Lathyrus latifolius* (see Figure 7.33), may be relied upon for producing extensive growth and numerous flowers in blue or white and is capable of finding support from surrounding plants, although for preference supply some beech twigs driven into the ground around it.

Figure 7.33 *White everlasting sweet pea* (Lathyrus latifolius), *hardy border perennial*

h. Growing specialist plants

Growing herbaceous perennials from seed allows the production of unusual plants which are not normally available on the market. It also presents a challenge for the experienced gardener and his or her husbandry skills. Seed may be bought from the seed houses or via a garden centre or some very specialist sources, such as the seed distributed annually by the RHS to its members. Seed of obscure plants may need a period of low temperatures (vernalisation, of spring, was formally defined by the French botanist P. Chouard in 1960 as "the acquisition or acceleration of the ability to flower by a chilling treatment", but the phenomenon was well known before that). This is done before sowing the seed as a means for overcoming dormancy (see Underpinning knowledge 3.2 and 5.2). Vernalisation is achieved by mixing the seed with some moist sand in a polythene bag and placing it in the salad tray of the refrigerator for 8 to 10 weeks.

Sow the seed as previously described (Chapter 4) and place it in a greenhouse. Alternatively, seed requiring vernalisation may be sown in a seed tray or pot and placed in a cold frame for the winter where chilling can take place. Monitor the container from time to time as mice and other vermin may take an interest in seed as a winter food source. Germination can be a slow process, and with very specialist seed it is worthwhile keeping the pots for at least 18 months, as 2 winters may be required in order that vernalisation is completed. Giving up on unusual plants too quickly is a mistake, as not infrequently, after "all hope has gone", strong and vigorous shoots appear!

Once germination takes place, prick out the seedlings into larger pots or modules filled with a suitable seed compost. Some plants will require acidic conditions, whereas others will demand alkaline compost (see Underpinning knowledge 2.2). This means that composts must be either devoid of lime (calcium carbonate) or contain larger quantities of it. Most composts will contain a sufficiency of plant nutrients, possibly in slow-release formulations (see Underpinning knowledge 4.4). Initially, place the pots in the greenhouse or polythene tunnel. Some perennials require cool growing conditions.

When the plants are large enough, plant them into a nursery bed in the open. The nursery bed is simply land which has been well dug over in advance with manure or compost added several months ahead of its use. Level the land with a rake and incorporate a general fertiliser, such as "GrowMore" at 30 g per square metre, which will suffice for most species. Some specialist plants may have different requirements. Gardeners working at that level of expertise should consult works of reference for information on the husbandry of a particular plant species.

The plants will grow for some months, possibly years, in the nursery bed, hence weed control will be essential. This is achieved by laying black plastic sheeting over the bed which stops weed growth and retains moisture around the roots. Plant out the seedlings from their pots in the normal way by making a sufficiently large hole with a trowel through the plastic sheet, planting to the depth of the seedling root ball and firming with the fingers (see Chapters 4, 5 and 6). Water the plants well, and spread slug bait around the seedlings. Larger predators such as rabbits, hares or deer may become a nuisance, in which case erect netting around the bed. Maintain the plants by careful monitoring, removing weeds as they appear and leaves or any other litter. Plastic sheets "age" because of degradation caused by ultra-violet (UV) light and cycles of hot and cold weather conditions and may need replacement. Remove flowers from the plants, thereby encouraging vigorous vegetative growth. After about 18 to 24 months, the plants can be planted out in the border. Plant out as described in Chapters 5 and 6. Enjoy the pleasure of having grown something unusual and possibly a plant that is classed as "difficult to grow". It is possible that there will be some surplus plants and the gardener can trade these with his or her developing peer group. Collaboration of this nature brings added enjoyment and health benefits derived from gardening.

Underpinning knowledge

7.1 Light and dark trigger flowering

Flowering is a response by plants to environmental conditions, and as a result they change from vegetative to sexually reproductive growth. This is a good example of the interaction between the genetic constitution (genotype, G) and the environment (E), changing the physical appearance (phenotype, P) of plants (see Underpinning knowledge 1.4, 3.1 and 3.2).

"Why do plants flower?" is a question that has long intrigued botanists and horticulturists, and this is partially discussed in Underpinning knowledge 6.2. Basically, plants flower as part of the sexual reproductive process, and hence as a means for perpetuating the species. A realisation that plants interact with their environment enabled the cultivation of an increasing range of flowering plants obtained by plant hunters from the early 16th century onwards. From the early 20th century, scientists began explaining what triggered changes from vegetative to reproductive growth in plants. This required technologies which permitted the culture of plants in programmed and precisely regulated growth chambers. In these, plants were exposed to artificially created seasonal patterns of light and temperature in controlled conditions which encouraged or discouraged flowering. This revealed that alternations of light and dark in a daily 24-hour cycle, and shifts in their duration replicated seasonal changes, inducing or suppressing growth and reproduction. The term "photoperiodism" was coined by W. W. Garner and H. A. Allard working in the USA in 1920, and describes the influence of seasonal lengths of day and night as regulated by light on flowering. Changes in day length are perceived by the leaves and then stimuli are transmitted via a growth hormone to growing points where flowering is initiated.

Three basic forms of flowering responses are identified:

1. Plants that flower when the dark period is longer than a critical minimum are known as "short-day plants". These will normally flower in the autumn and early spring.
2. Those that flower in response to dark periods shorter than a critical minimum are termed "long-day plants". These will normally flower in the late spring, summer and very early autumn seasons.

Plants in each of these categories originated from warm temperate, temperate and sub-Arctic or sub-Antarctic regions where there is substantial variation in the length of light and dark periods through the year.

3. A third group of plants, day-neutral types, flower regardless of day length. These evolved in sub-tropical and tropical zones where there is little or no yearly variation in the length of daylight.

Flowering responses to light and dark are summarised in Figure 7.34.

Initially, it was thought that the length of a light period controlled flowering. Now, it is recognised that it is the length of the dark period that is critical in determining the onset of flowering, as opposed to the length of the light period. Unravelling understanding of means by which flowering is controlled laid the foundations for the vast international flower industry which today supplies blooms round the world within and outside their normal reproductive periods.

Long-day plants require day lengths between 12 and 16 hour periods of of light; short-day types flower in response to light periods of less than 14 hours. Typical long-day plants include: clover, hollyhock, *Iris*, lettuce, spinach and radish; while short-day types include: *Chrysanthemum*, goldenrod, *Poinsettia*, soybean and many weed species. Day-neutral plants include peas, *Phaseolus* beans (runner and green beans), sweet corn and sunflower. The degree of reliance on day length stimuli varies with different species. In some plants, the requirement is absolute, these are known as qualitative or obligate flowering plants, for example cocklebur (*Xanthium strumarium*), a native plant used as a model plant by researchers. By comparison, some species respond to both short and long days and are known as quantitative or facultative flowering plants, for example the cereals, wheat and rye. Tropical plants may

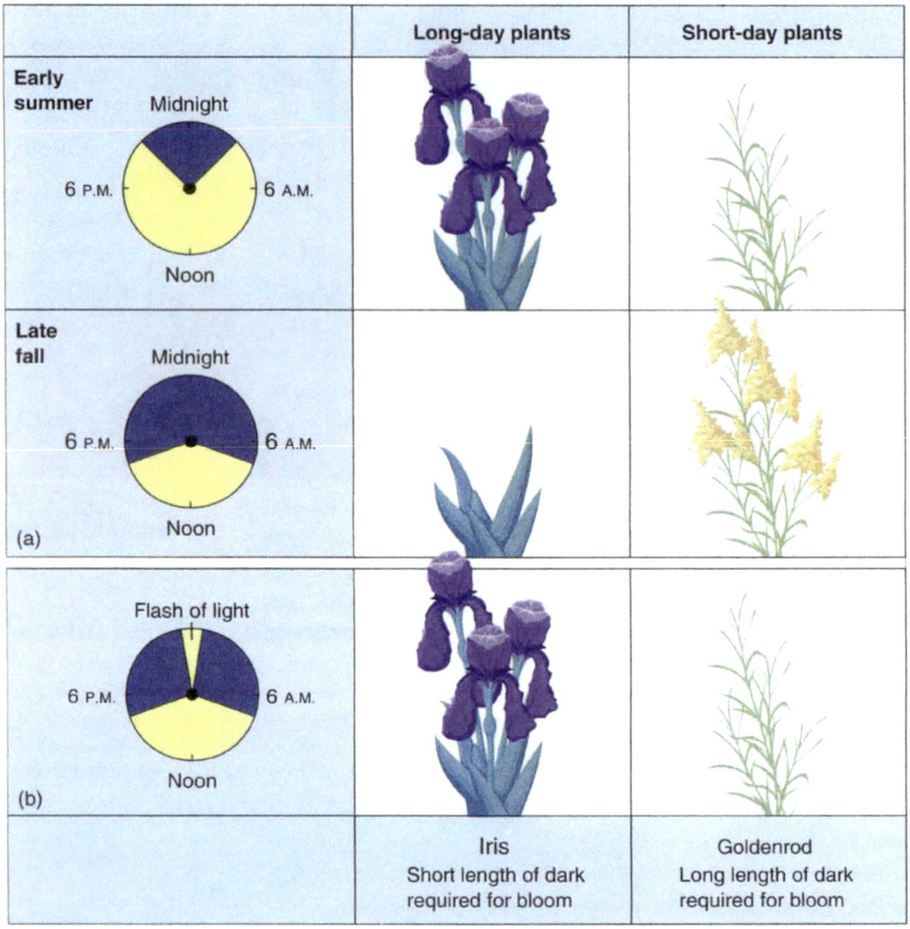

Figure 7.34 *Diagrammatic illustration of responses to day length in* Iris *spp. and goldenrod (*Solidago *spp.).*

have even more demanding requirements, for example some Indian grasses (*Sorghastrum nutans*) require two critical stimuli for flowering. These remain in a vegetative state when the day length is extended.

Early research focused on developing an understanding of the biology of short-day plants because the flower industry was especially interested in manipulating *Chrysanthemum* and *Poinsettia* crops. Critically, short-day plants require periods of darkness in the light period as the trigger for the onset of flowering. If they receive a very brief period of light interrupting that dark period, the flowering process delayed. From this, it was deduced that the effects of red light with a wavelength of 660 nanometres could be altered by a brief period of far-red light of wavelength 730 nanometres which then encouraged flowering. The light spectrum visible to the human eye has violet, short wavelengths at one end and red, long wavelengths at the other; beyond these are ultra-violet and infra-red wavelengths.

The regulation of flowering

Recognising the effects of dark periods on flowering brought an understanding that a plant pigment, known as phytochrome, is a key to the flowering process. Phytochrome is a single substance which exists in two states according to the wavelength of light which is received. One form reacts to red light wavelengths and the other to those in far-red part of the spectrum these are respectively termed P_r and P_{fr}. The P_{fr} state is biologically active and the P_r is inactive. In short-day plants, high levels of P_{fr} suppresses flowering; as this concentration declines in dark periods, there are increasing amounts of P_r and flowering develops. But a brief flash of red light of 660 nanometres will stop that development by restoring the level of P_{fr}.

Recent studies show that phytochrome-based systems for sensing light wavelengths are genetically ancient. The genes responsible for this reaction have been part of the genetic make-up for a considerable part of geological time. These probably evolved first in the early green algae which were the common ancestors of current marine and land plants. Phytochrome has several functions in plants. Seed germination, for example, is inhibited by far-red light and stimulated by red light. Chlorophyll absorbs red light, allowing the far-red wavelengths passage into and through the leaves. This inhibits seed germination in the developing green pods and seed capsules following flowering. It also suggests that when light passes through green foliage, red wavelengths are screened out and far-red light reaches the ground under trees. That stops seed germination when trees are in full foliage. The situation changes following leaf fall. In spring, before the leaf canopy has formed, red light reaches seeds below the trees and triggers germination. Seedling shoots are etiolated, pale and slender with an absence of chlorophyll when below ground, because they are growing in the absence of light (see Figure 7.35).

Once the shoots grow above soil level, red light speeds a change from the etiolated state into a green shoot (Chapter 1). Once this process has commenced, the emerging shoot transmits red wavelengths for several centimetres downwards into the growing shoot. Consequently, the whole shoot is primed with green chlorophyll and photosynthetically active when it emerges above the soil surface and that promotes more rapid growth.

Florigen is the name given to a growth hormone also associated with flowering. The existence of florigen was formally established in 2006, and it is now known to be a small globular protein. The production of florigen is genetically controlled and functions by terminating vegetative growth and inducing flowering responses.

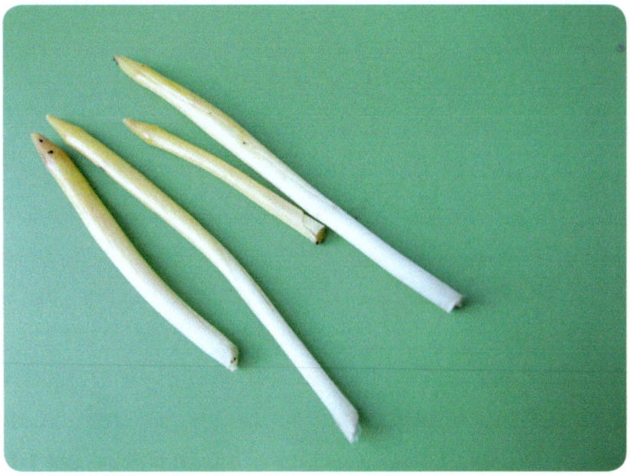

Figure 7.35 *Etiolated tulip shoots*

Alternative environmental responses

Other environmental factors are stimuli for flowering; of these, water is a powerful agent which induces flowering as in cacti (see Figure 7.36).

Figure 7.36 *Flowering in cactus* Rebutia *spp., stimulated into flowering by water*

In arid and desert areas, rain stimulates a very rapid growth response followed by luxuriant flowering such as that seen in the arid areas of North and South America, Australia's central desert and South Africa's Kirstenbosch National Park after the spring rains.

In dry environments, there may also be interactions with temperature. Dormancy may be regulated by temperature. Plants respond to falling temperatures by becoming acclimatised, increasing their potential for survival, but acclimatisation can be reversed by an onset of warmer conditions. Full dormancy develops after acclimatisation has increased plant resistance to low temperatures below a critical level. At that point, full dormancy is established and even a brief warming spell will not stimulate a growth response. This dual system of acclimatisation and dormancy protects plants which grow in northern and southern cool temperate zones from the effects of severe winter conditions and from making abrupt changes in their growth patterns (see Underpinning knowledge 5.2 and 5.4). During dormancy, the buds of deciduous trees and shrubs are encased in protective scale leaves. Since it is difficult for plants to take up water from frosted or frozen soil, dormancy in the buds prevents plants making aerial growth when the roots would be incapable of servicing the demands of foliage for water via a transpiration stream.

Dormancy is broken after a critical period of cold, which varies with different species depending on the location in which they evolved. Plants require a certain period of cold conditions at or below a particular temperature before satisfactory regrowth is triggered. In many trees and shrubs of temperate origins, bud burst and foliage growth can be preceded or followed quickly by flowering. Here, the flowering trigger is the change from dormancy and the onset of warming conditions, which allows expansion of the flower buds when they shed the protective scale leaves and come into flower (see Underpinning knowledge 5.1 and 5.2). The flower bud itself has developed in the preceding late summer and early autumn following or accompanying fruit development and maturation. When the leaves of these plants are shed, buds, which will open in the following season, become apparent.

During the winter period, perennial herbs are reduced to underground roots and rhizomes which are physically protected from cold and freezing temperature. Annual plants reside as seeds protected by a layer of soil or leaf mould. Seed can be protected from premature germination by an inhibitor which must be leached by water from the seed coat. Alternatively, the seed coat may physically crack and break open before the radicle and plumule can emerge. In some plants, seed germination only happens in cool conditions. This is termed "thermo-dormancy" and is a protection against germination during periods of hot, dry conditions when there is insufficient soil moisture and the temperatures are too high for satisfactory growth. A typical example is lettuce where temperatures greater than about 15 °C inhibit germination.

7.2 Adverse conditions

Shortages or excesses of environmental factors are causes of plant stress closely similar to the effects of pests and diseases.

Plants and animals suffer from stress disorders caused by adverse biological and environmental conditions. Adverse biological conditions may result, in plants, from the presence of a disease-causing organism, such as a pathogenic microbe or a pest such as an insect, and these are generally known as biotic stresses (see Underpinning knowledge 4.3). The environment normally provides plants with water, warmth, oxygen, carbon dioxide, light, nutrients and support (see Chapter 1). Depending on their geographical origins, plants have evolved with capabilities for utilising these resources within a certain range. Stresses result where physical atmospheric or soil conditions are above or below a normally acceptable range; these are termed "abiotic stresses". Because of their very diverse origins, plants used in gardens are individually capable of accepting and using resources over a wide range; compare, for example, hardy perennial plants with those of tropical origins.

Individual stresses

Water

Shortages of soil water cause drought at the roots, resulting in wilting of leaves and flowers. Conversely, excess soil water causes waterlogging, and the roots are suffocated because of a lack of oxygen, impairing respiration which results in wilting of leaves and flowers. In both cases, the servicing abilities of root systems are disrupted because the root hairs are very delicate single cells which are easily damaged by water shortages. Superficially, the symptoms of wilting resulting from both causes are similar. Some plants known as xerophytes, evolved in very dry and semi-desert areas, and are capable of withstanding considerable levels of drought but will be rapidly damaged by waterlogging. Others, evolved in areas where the soil is well supplied with water, known as mesophytes, are easily damaged by even limited drought or waterlogging.

Figure 7.37 *Effects of water stress and drought on the root and shoot size of* Viola *'Blackberry' plants. Left-hand plant well watered, right-hand plant watering restricted*

Drought stress in mesophytes has a severe effect on growth, especially in the roots (see Figure 7.37).

Temperature

Low temperatures cause chilling and eventually freezing damage in both aerial and root organs (see Figure 7.38).

The temperatures at which plants start suffering from chilling damage depends on their origins. Plants can be protected from cold damage in winter by wrapping in fleece, straw or "bubble polythene" (see Figure 7.39).

Plants from tropical zones are damaged by temperatures in the range of 5 to 10 °C. Those originating in northern or southern temperate or subarctic zones withstand temperatures below 0 °C. Conversely, excessive heat also causes stress damage and plants with Arctic or sub-Arctic origins are damaged by those within the normal range accepted by tropical ones.

Gases

Excesses or deficiencies of atmospheric gases cause abiotic stresses for foliage, flowers and roots. Aerial organs turn yellow, are disfigured and root growth ceases (see Figures 1.6 and 1.7). Nonetheless, altering the balance and concentrations of gases in packages containing harvested fruit and vegetables is used commercially as a way of extending their longevity in the sales chain from field to the consumer, a technique known as Modified Atmosphere Packaging

Figure 7.38 *Effects of low temperatures and frost, damaging the flowers and buds of* Viburnum farreri

Figure 7.39 *Protecting plants by wrapping in sheets of bubble polythene*

(MAP). Usually, the commercially changed atmospheres are accompanied by reduced temperatures. This cuts down the rates of metabolism, but if applied for too long then tissue breakdown develops. The result is an internal tissue damage, with the softening, yellowing and browning of fruit and vegetables such as apples, pears, tomatoes, cucumbers and peppers. Gases not normally present in the atmosphere, such as oxides of nitrogen and methane, cause substantial abiotic damage seen initially as areas of yellowing and death on leaves, stems, flowers and fruit. As concentrations rise or exposure is prolonged, damage becomes greater, eventually killing the plants.

Light

Plants are capable of using light of varying intensities. Those originating from the bottom of the canopy of tropical rainforests utilise light of low intensity and are damaged by, for example, bright sunshine. Such plants are valuable domestically as internal decoration or for landscaping inside buildings. Plants originating in areas of high-intensity sunshine such as mountainous regions utilise this intense radiation. Both types of plant are damaged by light with wavelengths beyond the visible spectrum, ultra-violet and infra-red, in excessive intensities for prolonged periods, which damages growth and reproduction.

Nutrients

Nutrient excesses cause stress at the root surfaces resulting in scorched roots and rapid death (see Underpinning knowledge 5.4). This is because the salt content or the acidity or alkalinity of the soil has exceeded acceptable concentrations (see Figures 2.50 and 2.52). Nutrients sprayed onto plant foliage at too high concentrations will similarly cause scorching and foliar death; this is particularly the case when young plants or seedlings are provided with liquid fertilisers. Deficiencies of nutrients disrupt plant metabolism resulting in a range of symptoms (see Figure 2.49).

Support

Soil or other substrates physically support plants as a result of their root penetration and spread (see Figure 1.3). Fibrous-rooted plants develop widely spreading root systems restricted to the upper surfaces of soil and substrates. These plants are often of relatively low in stature because the root system is physically incapable of supporting tall, stem extension growth. Plants with naturally deeply penetrating tap roots form trees and shrubs, some of which are capable of reaching very considerable heights, such as the forest giant trees. These heights are achievable because of a combination of deeply penetrating roots and widely spreading laterals reaching considerable distances. And the capabilities of the stem (tree trunk) for expansion in girth by the growth of annually produced meristematic cells in the cambial regions which produces a concentric cylindrical structure (see Figures 4.55 and 4.56).

Some trees, such as pines and beech, may lack deeply penetrating tap roots, especially when growing in very shallow sandy soils or where there are impermeable layers of compacted sub-soil or rock strata are close to the surface. Impermeable layers in the soil are known as "pans"; these may be natural in origin or arise because of continual cultivation at a uniform depth, producing "plough pans". Some plants compensate for natural pans by growing in groups or stands. Alternatively, some plants have evolved growth strategies which dispense with self-support and rely on climbing and twining on and around others. This provides the garden with useful species and varieties of *Clematis*, for example.

Support is disrupted by excessive wind velocities. Wind can damage young plants because the roots are insufficiently stable, causing desiccation and death. Wind-rocking of larger and more mature plants eventually causes the collapse of the plants, especially where rooting has been restricted. Where wind is a common phenomenon, it prunes and shapes plants, and this is a common sight especially in coastal areas. Wind transports fine soil particles which are abrasive, damaging emerging seedlings and transplants (see Figure 2.43). Salt (sodium chloride) carried by wind in coastal areas scorches and desiccates foliage, often killing tender plants. Species originating from windy maritime areas such as *Fuchsia* and *Olearia* are capable of withstanding far greater wind exposure and salt damage compared with lush and tender mesophytic plants.

Predicting stress damage

Predicting susceptibility to abiotic stresses or disorders requires an understanding of plant origins. This approach produced, for example, hardiness ratings indicating the capabilities for resisting low temperatures or susceptibility and resultant death. The American Department of Agriculture (USDA) developed a system of hardiness ratings which is used worldwide. More recently, the British Royal Horticultural Society (RHS) developed a similar system.

Some plants are capable of becoming acclimatised (acclimated) for particular abiotic stresses. Applying the stress at moderate levels or concentrations over a period of time increases tolerance for more severe conditions. The practical application of acclimatisation is achieved by providing winter protection such as wrapping plants in fleece in advance of more severe conditions.

Waterlogging is avoided by ensuring that soil is healthy and of high quality by regular cultivation and the additions of ample farmyard manure or composts. This improves the crumb structure, allowing water percolation, freedom for root development and ample aeration. Nutrient imbalances are avoided by careful application of fertilisers, ensuring that a quantity is applied which is compatible with plant requirements, but is not used in excessive quantities causing an accumulation of salts in the soil.

Wind damage is avoided by planting windbreaks, which in the larger garden should consist of a balance of suitable trees and shrubs which filter and slow the velocity of gusts. Artificial windbreaks constructed of plastic webbing held on posts can be used as an alternative while trees and shrubs are developing, but this detracts from the visual quality of a garden. Individual plants should be protected from wind damage by staking and tying very early in their development and supplementing this throughout the season. Supports vary: they may be bamboos canes, birch or boughs from other trees, metal stakes or netting.

Excessive quantities of atmospheric gases are an unlikely garden hazard. When these are present, pollution is probably resulting from industrial sources, requiring action from the relevant agency.

Monitoring the garden for the incidence of abiotic stresses requires frequent inspection of the plants for incipient symptoms. Drought causing wilting is very easily spotted and should be remedied immediately before damage becomes permanent. Frost damage is anticipated from meteorological forecasts, and when frosts are predicted, tender crops such as emerging potatoes are protected by increased earthing-up or by spreading wet newspaper and old rugs over the foliage (Chapter 2). On a longer-term basis, the use of cloches, glass or plastic, placed over the developing seedlings provides greater reassurance. These are especially valuable for overwintering crops like broad beans, winter peas or early spring plantings of tender legumes, lettuce and other salads. Difficulties arise, particularly with evergreen trees and shrubs which may not show symptoms of damage until some months or even years after the event, by which time remedies are too late.

Stress biology

Each stress causes broadly similar events in plant and animal cells. The disruption of cell membranes caused by low or high temperatures, wilting and other factors results in discharges of fluids followed by the collapse and death of leaves and flowers. Before the development of visible symptoms, cellular processes release damaging forms of "active" oxygen accompanied by the presence of specialised forms of enzyme. These processes causing damage in plant cells closely resemble the effects caused in animal cells. Stimulated by stress, cells release poisonous substances causing self-destruction. Much recent detailed research shows considerable parallels between pathogen induced diseases and abiotic stress symptoms. At the cellular and molecular levels, human cancer, tobacco-smoking and drought or cold damage in plants have biochemical similarities. Plants, animals and humans all require balanced diets, and where one factor is present in excess health declines, possibly irreparably.

The prevention of abiotic stresses by spraying with chemicals has been attempted. Anti-cryogenic, or cold-avoiding, sprays used ahead of periods of low temperatures achieves some protection from chilling and cold damage. But these materials require regular repeated applications because they are water soluble and easily washed off by rain. Some of these materials are available from garden centres and websites. Spraying with seaweed extracts has been shown to provide similar protection but again repeated applications are needed. Spraying water over crops, especially soft fruit, during periods of low temperatures is an effective measure. This makes use of the warming effect (latent heat of fusion) obtained when the water freezes which is often just sufficient to protect blossoms from damage. Nutrient and water stress are avoided by skilful husbandry and wind damage by planting windbreaks and the use of protective structures such as cloches and staking plants well before damage is likely. Useful indicators of local problems are gained from observing precautions used in by neighbours.

Achievements: have you understood and learnt?

At the end of the chapter, you should be able to:

1. Understand the differing types of herbaceous plant, their seasonality and reasons for their suitability as related to environmental conditions.

2. Know the equipment, facilities and husbandry required for sowing small-seeded herbaceous plants such a pansy in modular trays.
3. Understand that modular plants of herbaceous subjects such as pansy may be obtained directly from garden centres or mail-order companies and planted in the garden.
4. Understand how biennial herbaceous plants may be raised from seed and the provisions required for cultivating them including the use of nursery beds.
5. Know the processes and procedures required for planting out herbaceous biennial plants in their flowering positions.
6. Understand that some herbaceous plants are perennial and able to be grown in the garden for several seasons and their pre-planting, planting and post-planting husbandry.
7. Understand that for some herbaceous perennial subjects there may be a requirement for chilling the seed (scarification and vernalisation) before it may be grown and the conditions required for achieving this.
8. Understand the influence of differing lengths of daylight on the flowering periods of plants related to days length and that some plants do not react to this.
9. Understand that water and temperature may have an effect on the growing and flowering periods of plants.
10. Understand that plants and animals experience stress when the surrounding environmental conditions are more severe than is normally the case.
11. Know that stress reactions produce symptoms which closely resemble those caused by pest invasion and pathogen infection.
12. Know that there are several scales which identify the hardiness of garden plants against low temperatures, which enables gardeners to estimate the suitability of species for use in their conditions.
13. Understand that gardeners employ various means for avoiding stress damage to plants particularly guarding against damage by low temperatures in frosty conditions.

Definition of terms: *Describe* is able to show why a scientific principle is important; *Understand* applies scientific principles into the context of gardening; *Know* carries through scientific principles in practical gardening activities.

Chapter 8
Plant propagation

Introduction

Understanding and practising forms of asexual vegetative propagation increases the gardener's appreciation of plants as living entities and their relationships with the environment. It also brings considerable pleasure which comes from achieving the successful manipulation of biological processes. This understanding begins developing linkages between the practices of gardening and the science and arts of horticulture.

Vegetative plant propagation is one of the oldest gardeners' skills. It has been practised and refined since the earliest civilisations of the Babylonians and Chinese, who recognised their need for stable supplies of food and decorative plants. Sexual propagation using seed collected from cereal and other staple crops served early farmers and gardeners well. Probably in parallel, however, it was recognised that vegetative propagation produced more satisfactory results for some fruit and ornamental plants. This route conserved favoured characters such as yield, leaf shape, flower colour, fragrance, vigorous habits and resistance to pests, pathogens and environmental stresses (see Chapter 3).

The practical principles of vegetative propagation became well established by skilful gardeners. Its capacities and capabilities were brought to high degrees of perfection, firstly in Europe from the 16th century onwards, and latterly in North America from the 18th century. When new and unusual plants started arriving in Europe as a result of extensive geographical explorations and colonisations round the globe, they were multiplied in botanic gardens and landed estates. Similarly, substantial expertise was developed in Asia and Asia Minor which perfected the propagation of indigenous plants from much earlier eras.

From the start of the 20th century, horticultural scientists have unravelled many of the biological principles underlying successful vegetative propagation. Increasing knowledge of the formation and interaction of several growth-promoting substances (hormones) increased practical possibilities for the initiation and development of new roots and shoots on plants.

This chapter describes three examples of vegetative propagation: softwood, semi-ripe and hardwood cuttings using plants which respond by producing roots relatively speedily and also considers propagation by the division of a herbaceous perennial, rhizomatous *Iris* and of lily scales (leaves).

Garden practices

a. Softwood cuttings: the florist's zonal or regal *Pelargonium*

The florist's *Pelargonium* is one of the most popular of summer-flowering plants (see Chapter 7). It is used as specimen dot plants in borders, plantings *en masse*, and in tubs and urns as decoration for patios and piazzas or in large pots as internal decoration as described in Chapter 7.

The closely related small-flowered *Geranium* spp. make excellent vivid and long-lasting additions for herbaceous borders. These offer a wide variety of flower, leaf and foliar colours and shapes providing considerable opportunities for experimentation in garden design.

The equipment required for propagation includes:

1. Rooting-hormone powder: (rooting-hormone powders or liquids encourage root formation by stimulating cell division.). There are various proprietary formulations suitable for different types of cutting. The formulations differ in the types of hormone used (see Underpinning knowledge 8.1), and *Pelargonium* requires the type suitable for softwood cuttings. This is obtained from garden centres or mail-order suppliers. Place a small pile of rooting powder in a plastic saucer.
2. 7 to 8 cm diameter plastic pots and rooting compost.
3. Plastic saucer filled with water.
4. Stock plant of the florist's *Pelargonium* with plenty of side shoots; this is best obtained by setting aside one plant from a batch bought for bedding displays and grown on in a larger pot. Removing the flowering shoots encourages the development of vegetative side shoots which can be used as cuttings.
5. Gardener's knife with a curved blade about 6 cm in length and a sharpening stone. The knife is sharpened by running the edge of the blade down the sharpening stone at an angle of 45 °. Turn over the knife, running the reverse edge down the stone. The keenness of the blade is tested by gently feeling the blade with the thumb. It should have a sharp feel against the pad of the thumb. Always use the blade in movements away from the body not towards it (see Figure 8.1). When not in use, the blade should be folded away and the knife placed in a safe place.
6. First-aid kit for minor cuts.

Place the stock plant on a wooden bench at a convenient and comfortable working height (see Figure 8.2). This should allow the hands-free movement from the elbows without stretching or stooping, with the feet firmly placed at about 15 cm apart on level ground or concrete.

Figure 8.1 *Sharpening a gardener's knife*

Figure 8.2 Pelargonium *stock plant supplying cuttings*

Examine the stock plant and decide on the shoots which will most effectively supply softwood cuttings. Place the knife, rooting powder and water conveniently close for ease of working (see Figure 8.3).

Cuttings should be vigorously growing shoots about 4 to 5 cm long without signs of browning or ageing at the base. Cut cleanly through the shoot, taking the blade away from the body. Use the thumb as an anvil placed behind the shoot. Repeat this process until sufficient cuttings have been cut off (see Figure 8.4).

Each cutting should be straight and without any signs of pest and disease damage. Cut off the leaves from the lower 3 to 4 cm of the shoot; this reduces water stress in the cuttings. Ensure that no snags are left on the cutting where the leaves are removed as these will form openings for diseases such as grey mould (*Botrytis cinerea*). Wet the cutting with water from the saucer and dip the basal 1 cm into the pile of rooting powder (see Figure 8.5).

Figure 8.3 *Utensils used for propagation, rooting powder, knife and water*

In advance, fill each 11 cm diameter plastic pot with rooting compost up to within 2 cm of the rim. Make a hole in the centre of the compost and insert the treated cutting into it, gently

Figure 8.4 *Excised stem cuttings of* Pelargonium

Figure 8.5 *Stem cuttings of* Pelargonium *with the lower leaves removed and bases treated with hormone-rooting powder*

firming the surface (see Figure 8.6). Place all the pots in a large carrying tray. Water well using a can fitted with a fine rose, fully wetting the compost (see Figure 8.7). Either place the pots and cuttings onto a mist propagation bench which is heated electrically from beneath; or, fit a large polythene bag or document "polypocket" over each pot as a tent and stand it on a well-ventilated greenhouse bench (see Figure 8.8).

Mist propagation is a process by which cuttings either in containers or inserted into a layer of sand are retained in a humid environment by automatic intermittent spraying with a fine water mist until roots have formed. Rooting is sometimes accelerated by placing electrically heated cables in the sand.

The greenhouse should be covered in summer, with white shading preventing excessive temperatures from developing. These cuttings will root most effectively in a cool environment.

Where a mist bench is used rooting will take place in about 8 to 10 days. Plants placed in a polythene tent may take a little longer. The cuttings in either environment should be inspected on a daily basis. Leaves that fall away from the cuttings should be removed as these form further foci for grey mould (*B. cinerea*) infections. Plants in polythene tents require removal of the bag each day and turning it inside out, removing condensation.

The development of roots is tested by gently slightly pulling the cutting, when roots have emerged then there will be a slight resistance. At that stage, the cuttings are becoming independent plants.

Figure 8.6 *Stem cuttings inserted into pots of compost*

Figure 8.7 *Pots containing stem cuttings of* Pelargonium *placed on the greenhouse benching and watered well*

Figure 8.8 *Pots and cuttings covered with plastic "document pockets"*

Figure 8.9 *Rooted cuttings of* Pelargonium

Rooted plants grown on mist benches will need a period of weaning when the frequency of misting is reduced in steps until it ceases completely. Plants rooted under polythene tents should have the sides of the bags opened and gradually lifted until they are capable of existing without them. Well-rooted plants (see Figures 8.9 and 8.10) should be watered with a can fitted with a fine rose every few days.

Re-potting well-rooted plants into a potting compost will provide the new plants with additional nutrients and increase their rate of growth. New pots, 11 cm diameter, should be half filled with a general-purpose potting compost. The rooted plant eased out of the old pot and held over a new one, replacement compost should be dribbled around the new roots and up to the neck of the plant. Firm the compost with your fingers.

Figure 8.10 *Adventitious root systems produced by new* Pelargonium *plants*

The plants should be placed back onto the greenhouse bench and watered using a can fitted with a fine rose and monitored regularly. Since the plants have been propagated in the shade, they will be susceptible to sun scorch. This is especially the case for those propagated using a mist bench, as this removes the protective wax from the surface of the cuticle of the leaf. Normal growth should commence in 7 to 10 days, with each plant producing a new crop of vegetative shoots and flower buds. This provides a further batch of florist's *Pelargoniums* for use in borders or in pots and urns.

b. Semi-ripe hardwood cuttings: for example, *Hebe salicifolia*

The species and botanical varieties of *Hebe* are very decorative and robust shrubs for the garden, some are trees reaching 20 m in height. Most *Hebe* spp. suitable for the garden vary from more or less prostrate to 2 or 3 m in height. Their flowers are very attractive for bees. Propagation from semi-ripe hardwood cuttings is relatively straightforward and best accomplished in the early to mid-summer period when the plants are in blossom or just emerging from it (see Figure 8.11).

The requirements for making cuttings are similar to those used for softwood cuttings.

Cut a bough from a large *H. salicifolia* plant. Place it on a firm bench top which has been covered in some old newspapers or a hessian sack. Identify vigorous new shoots which may be used as cuttings. Cut off 10 or 15 cuttings which are about 8 to 10 cm long using a sharp gardener's knife or pair of secateurs (see Figure 8.12).

Figure 8.11 Hebe *var. 'Autumn Glory', suitable for semi-hardwood stem cuttings*

Figure 8.12 Hebe *var. 'Autumn Glory' shoots detached, ready for trimming into stem cuttings*

Remove all the lower leaves from these shoots up to about 3 cm from the tip and dip their bases in water and then into hormonal rooting powder (see Figure 8.13). Insert three cuttings into each pot of compost, placing them around the rim of a pot (see Figure 8.14). As previously with the softwood cuttings, water the pots well. Put them either on a mist bench or cover each with a polythene bag or document "polypocket" and stand them on the greenhouse benching (see Figure 8.15). The follow-up monitoring and weaning follows a similar pattern to that used for softwood cuttings. Roots will develop over about the next three weeks and the plants become capable of independent growth (see Figures 8.16 and 8.17).

Figure 8.13 Hebe *var. 'Autumn Glory' stem cuttings showing lower leaves removed and bases treated with hormone-rooting powder*

Figure 8.14 Hebe *var. 'Autumn Glory' cuttings inserted into pots of compost on the greenhouse bench and watered well*

Figure 8.15 Hebe *var. 'Autumn Glory' cuttings and pots covered with a plastic "document pocket"*

Figure 8.16 Hebe *var. 'Autumn Glory' cutting producing roots into the compost when taken from the pot*

Figure 8.17 *Adventitious root system produced from a cutting of* Hebe *var. 'Autumn Glory'*

Figure 8.18 Hebe *var. 'Autumn Glory' plant 9 months after vegetative propagation*

Re-pot the rooted cuttings individually into 11 cm pots containing general-purpose compost. These will become 10 or 15 new independent *Hebe* plants.

As growth develops, remove the apical growing point by simply pinching out the tip between the thumb and forefinger. This removes the apical dominance in the plant and encourages growth from the side shoots. The result will be a bushy, well-balanced flowering plant after about 9 months' growth (see Figure 8.18).

Hebe spp. are hardy shrubs so that the plants should be removed from the greenhouse as soon as vigorous growth develops. Either place these plants in the border, or trade them with neighbours and fellow gardeners for alternative border plants. Considerable enjoyment is gained from exchanging the products of propagation with fellow enthusiasts.

c. Hardwood cuttings: for example, blackcurrant *Ribes* sp.

Growing blackcurrants as a soft-fruit crop is described in Chapter 5. Each winter when the bushes are fully dormant, they will require light pruning, which maintains their shape and ensures an adequate air flow which discourages disease development. Pruning results in shoots which can be used as hardwood cuttings. Select 8 to 10 robust shoots, each about 15 to 20 cm long (see Figure 8.19).

Ensure that each is free from signs of pests and diseases. Re-trim the base of each shoot with either a very sharp knife or pair of secateurs, producing a cleanly made sloping cut with no snags. Ragged cuts are more easily invaded by fungi such as grey mould (*Botrytis cinerea*). Dip the base of each cutting in water, and plunge the basal 2 or 3 cm of the shoot into rooting-hormone powder suitable for hardwood cuttings obtained from a garden centre or mail-order supplier. A small amount of powder should have been previously decanted into a suitable vessel such as an old enamel mug, if working outside, which can be easily washed afterwards.

Insert the treated cuttings into a row in the nursery bed. This ground should have been previously dug and cultivated producing a piece of soil which is fertile, well drained (see Chapter 2) and sheltered from excessive winds and frosts. Blackcurrants resist freezing temperatures quite adequately, but frost pockets concentrate cold most damagingly when growth is restarting, which injures or kills newly emerging leaves and shoots. Frost pockets are areas of low-lying ground into which freezing air accumulates, and this can damage even normally frost-tolerant plants.

Figure 8.19 *Hardwood cuttings of blackcurrant suitable for inserting into a nursery bed in the garden*

Mark out a straight row with the garden line and push each cutting into the soil to a depth of about 10 cm with about 20 cm between each shoot. Monitor the cuttings, ensuring that they do not attract the attention of vermin such as mice. New early spring growth can be an interesting food source when others are scarce.

Test each cutting for the development of roots as early-spring soil temperatures rise above about 5 °C. That is the point when root growth restarts; in the south of England this happens in mid to late February, while in northern Scotland it may be delayed until late April or early May. Each cutting will offer slight resistance when gently pulled as roots form (see Figure 8.20) and buds on the stem commence growth, showing green colouration as leaves start emerging from between the brown scales that have protected them during winter. During periods of dry weather, provide water for the rooted cutting. Green shoots will produce extension growth with the emergence of leaves. When established, these plants will produce extensive root systems whicht are apparent in the following spring even before the foliage appears (see Figures 8.21 and 8.22)

Figure 8.20 *Adventitious root systems produced from a hardwood cutting of blackcurrant growing in a nursery bed in the garden*

The new plants may now be planted into their growing and fruiting stations. Place a digging fork underneath each new plant, gently easing it from the soil. The base of

Figure 8.21 *Adventitious root system produced from a blackcurrant cutting after 12 months' growth*

Figure 8.22 *Close-up of the adventitious root system produced from a blackcurrant cutting after 12 months' growth, showing extensive root hair formation*

the original cutting will have become a mass of healthy white roots. Plant the new bushes as described in Chapter 5. Those which are excess to requirements may be potted into 15 cm diameter pots of general purpose compost (Chapter 7). Place a split bamboo cane into the pot alongside the blackcurrant plant tying it with 4-ply garden twine. Place each pot outside in a sheltered space or in an open cold frame. Water the plants well with a can which is fitted with a coarse rose and monitor them at intervals. These plants can be traded with friends and neighbours or gifted for charity sales.

d. Plant division: herbaceous perennials

Herbaceous perennials in particular are increased by division. This has the double benefit of rejuvenating ageing plants which are several years old and as a result are producing a declining display and permits cleaning root clumps of invasive perennial weeds such as ground elder. A convenient subject is the border *Campanula*, a bell-flowered plant useful as low ground-cover with flower formed on stems 20 cm tall. It is valuable when planted towards the front of the border flowering in early June.

Campanula plants are best divided in the early March when they have just produced signs of spring growth showing leaves and some shoots. This is when the borders are being readied for late-spring displays by forking-over, weeds removed and mulches replaced. Division can be accomplished on the lawn adjacent to the border as described in Chapter 7, see also Chapter 6 for the division of *Agapanthus* plants.

e. Rhizomatous *Iris* spp.

Rhizomatous *Iris*, especially *I. germanica*, the bearded (or flag) irises, are some of the most treasured garden plants (see Figures 6.22 and 6.23). Individually, their flowering periods are short, lasting a couple of weeks but by selecting a suitable range of varieties a spread of flowering over 6 to 8 weeks can be achieved as illustrated in Chapter 6.

They have a variation of heights through small, medium and very tall flower spikes; consequently, a whole section of a border might be used for them. The flower spikes carry 5 or 6 single flowers with an enormous spread of colour, from almost pure white through deep purple to nearly black, varying between the standard (upright) and fall (downward-curving) petals of differing colours. Flowering irises make majestic statements in the border.

Irises of this type originate from dry, arid, warm temperate zones in areas of limited soil depth and low nutrient content. The rhizome is a swollen stem which stores resources, a sink. Feeder roots, which both anchor the plant and seek out nutrients and water, develop laterally penetrating the upper soil surface on which the rhizome sits exposed to light (see Figure 8.23).

This habit renders rhizomes vulnerable to damage from forks and feet when the borders are being cultivated. Rhizome growth extends from the tip and older portions dry out and die. As plants age, they form expanding clumps with new growth around the periphery, which may be invaded with perennial weeds such as ground elder (*Aegopodium podagraria*) (see

Figure 4.62). Consequently, as with *Campanula* plants, they should be lifted and divided every few years, which encourages new growth and removes invading weeds.

Prune the flower stalks down towards the rhizome surface, leaving a stub about 2 to 4 cm long, and cut the leaves down removing half their length. Place a fork under the clump of rhizomes and ease it from the soil. Large clumps may require easing with a fork around their entire periphery. Once out of the ground, place the clump on a piece of stout plastic sheeting or an old hessian sack, and move it to a bench in the garden shed. Examine the rhizome carefully determining how many off-shoots may be obtained from the main leafless older rhizome.

Cut off these offsets with a sharp gardener's knife taking care that they retain sufficient roots and discard the old rhizome (see Figure 8.24).

Each cut surface should be coated with yellow flowers of sulphur (powdered yellow sulphur) obtained from a garden centres or mail-order supplier. Flowers of sulphur, finely powdered yellow sulphur, is obtained from brimstone rock formed naturally near volcanos, it has been used for several millennia and is referred to in the Bible. This is a disinfectant which helps prevent rotting at the cut surfaces. Considerable numbers of cuttings may be obtained from one rhizome (see Figure 8.25).

Figure 8.24 Iris *rhizome divided for propagation*

Figure 8.23 Iris *rhizome suitable for propagation by division*

Figure 8.25 *Tray containing* Iris *rhizomes that have been divided*

These new plants require a couple of years' growth before they will flower and during that period should be grown in a nursery bed or in large pots (30 cm diameter). Plant them into open ground which has been dug and cultivated and is well drained, which encourages further root growth. Using a trowel insert each cutting into a shallow planting station. Once the rhizome is planted, pull soil around it, but do not cover it completely, the surface should be visible above the soil surface; firm the soil with your fingers. Water well with a can fitted with a coarse rose. No extra fertiliser should be needed, but sprinkle some slug bait around the rhizomes.

Figure 8.26 Iris *rhizome divided, planted into a pot of compost and watered well*

Alternatively, the rhizome cuttings may be potted-up using a general-purpose compost and placed in a cold frame (see Figure 8.26). Water well for the first 10 to 14 days. The rhizomes will start forming vigorous root systems quite quickly (see Figure 8.27).

Protect the pots from excessive rain but water in dry spells. In mild autumn weather, these plants will form buds and possibly some new leaves. Monitor the rhizomes either in the open ground or in pots through the winter, preventing them from being covered by fallen tree leaves or other detritus so that they remain dry. Keep the ground and pots free from weeds and in spring water in dry periods. Maintain monitoring growth through the summer, remove senescent foliage in the autumn and keep the plants clean and free from weeds. These plants are unlikely to flower for another 12 months. This extended period of cultivation over 2 seasons means that pot-grown plants require considerable monitoring in order that they thrive.

Figure 8.27 *Adventitious roots and side shoots produced from propagated* Iris *rhizomes*

The following late spring and early summer is when the open ground or potted plants may reach flowering size. The rhizomes will recommence growth in this second spring in the nursery bed and at that time may be planted back into the flower garden (see Figure 8.28).

Lift the rhizomes gently with a fork and place them in a bucket of water, ensuring that

Figure 8.28 *Propagated* Iris *rhizomes planted in a border*

the roots remain moist or remove them from the pots. Plant as quickly as possible. Place each plant in its flowering position in the *Iris* or herbaceous border. Plant shallowly in a hole made with a trowel. The entire rhizome with its roots should fit snuggly into the hole, place soil round the rhizome. Cover the roots but not the rhizome surface, gently shake the rhizome, settling it into the hole and firm carefully with the fingers. It is important that the upper surface of the rhizome is exposed to light as that encourages the development of flowers. Water well with a can fitted with a coarse rose and sprinkle some slug bait round the rhizomes.

Medium to tall flowering irises will require support for the flower spikes. Drive sufficiently long bamboo canes alongside the rhizomes and tie-in flower spikes as they emerge. Sprinkle fertiliser containing potash (K) and phosphate (P) around the rhizomes as a stimulant for root and flower formation. Adding nitrogen (N) can be counterproductive since it stimulates leaf growth at the expense of flower formation. The subsequent lush foliage will be susceptible to damage from pests and diseases. After flowering reduce the spike down to a 2 to 4 cm stub, removing senescent foliage, otherwise this will encourage a damp microclimate and offer habitats for pests and diseases.

f. Bulb propagation: scales (leaves) of *Lilium*

Bulbs are composed of a series of scales which are modified leaves with buds set on a basal plate as described in Chapter 6. Bulbs may be propagated from these leaf scales. Most bulbs are rather small and their propagation using leaf scales requires specialised equipment and facilities. Some larger bulbs such as the lily have larger scales and may be propagated from them.

In early spring, purchase a large lily bulb from the garden centre or a specialist mail-order supplier. The large, very colourful and highly scented varieties are ideal. Propagation requires:

1. seed tray filled with general-purpose compost,
2. sharp knife,
3. rooting-hormone powder suitable for softwood cuttings. Decant some into a small plastic saucer.

Place the bulb on a bench and cut off the older scales, particularly ones which are pock-marked or diseased from around its outside. Cut off undamaged scales (see Figure 8.29), moisten their bases with water and dip them into the rooting-hormone powder.

Plant them out in neat regular rows in the seed tray with 2 to 4 cm spacing between each scale (see Figure 8.30).

Figure 8.29 *Lily bulb scale leaves detached from a bulb for propagation*

Figure 8.30 *Lily bulb scale leaves inserted into compost in a plastic seed tray filled with compost and watered well*

Only the basal 0.5 cm of the scale should be inserted into the compost. Full trays should be placed on the greenhouse benching and watered well from a can with a coarse rose. Place each tray inside a large polythene bag. Alternatively, place the tray on a mist bench or in a warm environment such as an airing cupboard. Monitor the trays in each environment, preventing the compost from becoming waterlogged. Shake condensation out from the polythene bag as a daily routine and turn the bag inside out. After about 3 weeks, there should be signs of shoot growth around the base of the scale. Very tiny bulbils, small leaves and delicate roots will form (see Figure 8.31). Quickly remove each growing scale very gently with the bulbil, leaves and roots and pot separately into a 7 cm diameter pot filled with compost.

Make a small hole in the centre of the compost with a pointed bamboo stick and gently place the scale with the bulbil, roots and leaves into the hole drawing compost around it. The developing bulbil and leaves should protrude above the compost. Stand

Figure 8.31 *Lily scales producing new bulbils after propagation*

Figure 8.32 *Comparison of well-developed lily bulbils growing from leaf scales, with immature and flowering size bulbs*

the pots on the greenhouse benching and keep the compost moist but not waterlogged monitoring regularly. By late spring or early summer, these plants should have made significant growth which can be compared with immature and flowering size plants (see Figure 8.32).

Plant the bulbils out into a nursery bed using land which has been dug and fertilised where there is a good friable tilth. Follow the practices described in Chapters 3 and 4 for planting out young growing plants. The lily leaves should appear above soil level with the developing bulb below it. Sprinkle slug bait around these plants and water well using a can fitted with a coarse rose. Apply some general-purpose "GrowMore" fertiliser at half rate, i.e. 15 g per square metre. Vermin may be troublesome, since these bulbs contain sugars because they are sink organs; consequently, it may be necessary to set some mouse traps. Experience suggests that fragments of chocolate or coatings of peanut butter are as attractive, if not more so, than the traditional Cheddar cheese. Grow the bulbs in the nursery bed for at least two growing seasons at the end of which flowers may develop. "Nursery beds" are a gardeners' term for areas of land in which young developing plants are grown before being moved into their final positions. The nursery bed should be kept weed-free; additional slug bait and fertiliser will be required as the plants grow. Watering may be needed in dry seasons.

The bulbs should flower first in the nursery bed and only those with the finest colours and most attractive growth are retained. These may be used in the herbaceous borders. Lift the bulbs in the autumn and plant directly into the border in accordance with the planting design. Plant as described in Chapter 6. Alternatively, lift the bulbs in the autumn, drying them, place them in a well-aerated polypropylene sack, and suspend it from the roof trusses of an airy frost-proof garage or shed until spring-time. Lily bulbs make good additions to herbaceous borders, adding colour, height and considerable scent, especially on a warm summer evening.

Underpinning knowledge

8.1 Encouraging rooting

Artificially encouraging the initiation and development of roots is the basis of asexual plant propagation which has been referred to throughout this book. This requires their stimulation by the use of growth-regulating hormones and subsequent careful culture until the propagule has become a fully independent plant. Propagule is the term applied to detached plant parts prior to the formation of roots.

The anatomy of plants, particularly stems, is described in Underpinning knowledge 5.1. Asexual propagation with softwood, semi-ripe and hardwood stem cuttings encourages the capabilities for division and multiplication of the stem parenchyma and the vascular cambium cells. The processes of taking the cuttings wounds the base of these stems and that initiates wound-healing processes and the subsequent production of adventitious roots. Some plants such as willow (*Salix* spp.) or the winter-flowering jasmine (*Jasminum nudiflorum*) have pre-formed root initials within the stem and consequently are straightforward subjects for propagation and are natural examples of adventitious root formation (see Figures 8.33 and 8.34).

Other plants are more difficult and form a mass of undifferentiated cells known as callus, from which root formation is encouraged by the application of rooting-hormone powders. Each cutting relies on energy which is stored in the tissues and is mobilised during root formation or on its capture from sunlight during propagation. Softwood cuttings such as the *Pelargonium* retain green tissues in the leaves and stem and consequently photosynthesis and energy capture from sunlight continues during the rooting process. Semi-ripe cuttings such as the *Hebe* retain some of the more active leaves at the top of the cutting while those which are older and possibly commencing senescence are removed. There are also stores of carbohydrate within the stem cells which the cutting may utilise. Hardwood cuttings such as the blackcurrant rely on starch stored within the stem for energy supplies during root formation and subsequent bud break and leaf formation.

Root-stimulating substances

Stimulating root development within the tissues of these types of cutting requires a trigger or messenger. These are known as growth-regulating substances (hormones). Several groups of growth-regulators have

Figure 8.33 *Incipient adventitious root formation on the stem of* Jasminum nudiflorum

Figure 8.34 *Well-developed adventitious root system produced on a stem of* Jasminum nudiflorum

been identified in flowering plants. The key group involved with the stimulation of cell division and root formation are the auxins. Research from the 1920s onwards identified these materials, their functions and more recently the genetic systems by which they are inherited and triggered into action. The Dutchman Frits Warmolt Went recognised the functions of auxins in 1928.

When present in raised amounts in stems due to either natural causes or because of applications of synthetic auxins when propagating cuttings, these substances cause the "dedifferentiation" of parenchyma cells, which then give rise to emerging root initials. As these expand, they rupture through the stem tissues and emerge as adventitious roots which have vascular connections *in situ* linking with the xylem and phloem of the cutting and converting it to a free-living plant.

Auxin levels tend to rise naturally in the spring as growth recommences after winter and gradually decline through to autumn. Consequently, softwood cutting may root in the spring without the aid of rooting powder. Auxins tend to be in higher concentrations in the apical buds, leaves and shoots; hence juvenile material also roots more effectively than more mature tissues. This is one reason why stock plants are often "cut back", preventing them from flowering and encouraging the production of juvenile shoots which will root more quickly and effectively. These processes are replicated in research laboratories and commercially with plants that are difficult to root or where it is necessary that propagation is undertaken in sterile conditions, as, for example, when freedom from viruses is required as with soft fruit (see Chapter 5). Plants are grown from small pieces of undifferentiated tissue, callus, in the process of tissue culture which causes the formation of microplants (see Figures 8.35 and 8.36).

Figure 8.35 *Tissue culture (micropropagation) of young plants grown on agar supplemented with nutrients and growth hormones*

Figure 8.36 *Rooting of micropropagules in agar media just prior to weaning*

Growth is encouraged by placing small pieces of callus onto an inert medium known as agar which contains specialised recipes of chemicals which research has shown will stimulate the multiplication and differentiation of cells, eventually forming small new plants.

Weaning and plant establishment

Softwood, semi-ripe wood cuttings and the products of micropropagation are encouraged into rooting and subsequent growth by either placing them on a mist bench or covering with clear polythene bags or tents. The objective of either technique is conservation of a humid atmosphere around the cuttings so that they do not dry out. Effectively, the process of transpiration is blocked because the high humidity physically prevents the release of water vapour through the stomata of leaves, the softwood stem or micropropagules. The transpiration stream in the cutting is put into abeyance.

Mist benches operate by intermittently spraying a fine mist of water droplets over the cuttings from specially constructed nozzles (see Figure 8.37). This is regulated by an "artificial leaf" which, when wet, prevents an electrical current from flowing (see Figures 8.38 and 8.39).

Once the electronic leaf dries, current will flow, completing a circuit which then causes a mist of small water droplets to be emitted rewetting the sensor and the leaves. The frequency of misting can be set at a predetermined frequency by timers. Warming the root zone using electrically heated cables speeds up the rooting process (see Figure 8.40).

As rooting starts, the frequency of misting is reduced such that when it is fully completed the humidity around the cuttings which are now rooted plants returns to normal ambient relative humidity values. This process is termed "weaning" because the dependent cutting very gradually becomes self-supporting. It is then weaned from dependency. The quicker this can be achieved, the better because misting leaches nutrients from the leaves and consequently, if continued for too long the cutting becomes deficient in essential nutrients. Full automation of mist propagation using timers and sensors (see Figure 8.41).

Figure 8.37 *Mist bench atomiser nozzle*

Figure 8.38 *Electronic leaf that controls misting*

Figure 8.39 *Electronic leaf and cabling*

Figure 8.40 *Electrically heated warming cables inserted into sand or gravel of the mist bench*

Figure 8.41 *Illustration of a propagation bench with controlling timer box and mist nozzle mounted above the bench*

Figure 8.42 *Weaning plants commercially using large-scale misting*

A similar process is carried through with cuttings rooted under plastic bags or tents. Each day, these are lifted and the accumulated condensate is removed as this would encourage fungal organisms which could lead to rotting of the cuttings. The cuttings are then treated with a fine water spray and the tent is replaced. Gradually, as rooting takes place, the tent can be left off the cuttings for increasingly longer periods. This weans the rooting cuttings off dependence on the humidity under the tent and is comparable with weaning plants from a mist bench. In commerce where very large and highly valuable batches of plants are being produced, entire glasshouses can be filled with mist, which forms a fog (see Figure 8.42). In the garden, as soon as the cuttings have become self-supporting, they should be re-potted into a standard growing compost and placed out on the greenhouse benching. These plants need close monitoring for some time, as they are not fully weaned until an expanding root system has developed in the new pot.

Achievements: have you understood and learnt?

At the end of the chapter, you should be able to:

1. Describe why asexual propagation is an essential aspect of gardening.
2. Understand the sources of softwood, semi-ripe and hardwood stem cuttings.
3. Know how softwood stem cuttings are taken from a mother stock plant, prepared, held in humid conditions until rooting commences, weaned and potted as independently growing plants.
4. Know how semi-ripe wood stem cuttings are taken from suitable plants growing in the garden, prepared, held in humid conditions until rooting commences and potted as independently growing plants.
5. Know how hardwood cuttings are taken from suitable plants growing in the garden, prepared and inserted into a nursery bed and husbanded through winter conditions until green shoots develop and roots are produced in the following spring forming independently growing plants.
6. Know how herbaceous perennial plants may be multiplied asexually by the division of larger existing garden plants and their subsequent cultivation.
7. Know how the rhizomatous stems of bearded irises are divided in order that more plants may be obtained.
8. Understand the structure of lily bulbs and the opportunities for their asexual multiplication using the leaf scales and the subsequent husbandry required.
9. Describe the process by which roots are formed inside stem tissues and the roles of growth-regulating substances in stimulating these events.
10. Understand why a period of weaning is important in the cultivation of recently rooted new plants.

Definition of terms: *Describe* is able to show why a scientific principle is important; *Understand* applies scientific principles into the context of gardening; *Know* carries through scientific principles in practical gardening activities.

Tailpiece
Successes and failures

Background

Gardening is both a hobby and profession which uses and adapts biological processes; it combines both the arts and sciences, involving an intimate relationship between people and plants.

Biological processes are inherently variable because they deal with living organisms interacting with their surrounding environment, which itself varies with daily changes in weather and the seasons.

Inevitably, therefore, gardeners will have successes and failures, in that they are no different to the experiences of highly skilled, scientifically and technically able professional horticulturists. Gardening and horticulture are both continuous learning processes. Even the most able professional horticulturists never stop learning because biological processes always retain a few surprises. The relationship between gardening and horticulture may be summarised by the following: "gardening is to horticulture as the theatre is to literature". Gardening contains no "quick fixes", trying to take short cuts is usually doomed to failure.

The key take-home message from this book is that successes should be celebrated as triumphs and be relished, while failures are platforms for future learning and improvement. Turn failures into successes by discussion and analysis with others. Blame, attributed to plants, seeds, animals, pests and diseases, other people and the gardener him- or herself does not exist in gardening or horticulture. Neither gardening nor horticulture are "blame games"!

There is no necessarily single "correct" way of achieving success in the garden, but some pathways and techniques are easier, better suited, more disciplined and achieve results more quickly.

Gardening successes stem from:

- Recognising failures and learning from them.
- Close attention to the welfare and wellbeing of plants.

- Making regular tours of the garden watching for the incipient signs of developing problems.
- The use of good-quality tools and sundries.
- Purchasing seed and plants from reliable sources, especially those with qualified and knowledgeable staff.
- Meeting and discussing the garden with neighbours and fellow enthusiasts who have experienced the locality's environment for some years.
- Gaining information and knowledge from national and local sources.
- Developing garden soil which is healthy and of high quality by careful husbandry over a period of years.
- Sowing and planting crops in their appropriate seasons while experimenting with early and out-of-season plants.
- Ensuring that perennial plants are given care and attention as they enter dormancy, and during early stages of development in the spring.
- Eradicating weeds, pests and diseases at the first signs of invasion and infection.
- Harvesting crops in the early stages of maturity when quality and flavour are at their peak.

Failures usually result from:

- Purchasing apparently cheap tools, sundries, seeds and plants—curiously, in the long run, these rarely turn out as "cheaper".
- Not eradicating perennial weeds or dealing with long-term pests and pathogens.
- Not treating soil as a living entity, and stifling its life by causing compaction and waterlogging.
- Assuming that gardens require attention only during spring, summer and early autumn, whereas late autumn and winter-time tasks lay the foundations for success during the next and subsequent growing seasons—gardening does not have "a closed season".
- Not gaining background knowledge and information concerning why and how plants grow and thrive.

Sources of further reading and information

Sources of further information and knowledge are provided in the following lists. Some of the books cited are modern treatments; others are older but still contain relevant material. The lists of websites and apps offer current sources of more information and a short list of suppliers. None of these lists is fully comprehensive; the absence of particular sources does not represent a judgement of quality or value. Comments on the topics covered in books are purely personal considerations in an attempt to convey content, again without any critical assessment of quality or value.

Both these books are well-conceived texts suited for advanced students and professional horticulturists.

Adams, C., Early, M., Brook, J. & Bamford, K. (2015). *Principles of Horticulture: Level 2*, Routledge, Abingdon, Oxon.

Adams, C., Early, M., Brook, J. & Bamford, K. (2015). *Principles of Horticulture: Level 3*, Routledge, Abingdon, Oxon.

A survey of the biology and functioning of plants, this book accompanied the television series.

Attenborough, D. (1994). *The Private Life of Plants*. BBC Books, London.

These RHS books are written by authoritative practitioners.

Baker, H. (1994). *Growing Fruit: The Royal Horticultural Society's Guide to Practical Gardening*. Octopus Publishing, London.

Biggs, T. (1994). *Growing Vegetables: The Royal Horticultural Society's Guide to Practical Gardening*. Octopus Publishing, London.

Brickell, C. (ed.) (2016). *The Royal Horticultural Society's Encyclopedia of Garden Plants*. Dorling Kindersley, London.

These three texts are produced by one examination board and provide students with information attuned to their modules; other examination boards will have similar texts.

Burrows, C., Faudemer, K., Howe, E., McGarry, C., Pattison, S. & Walsh, S. (2016). *GCSE AQA Biology, for Grade 9–1 Course*. Coordination Group Publications (CGP), London.

Burrows, C., Faudemer, K., Kordan, R., Lindle, C., Marshall, R., McGarry, C., Pattison, S. Plowman, C., Rogers, R., Simson, C. & Thompson, H. (2015). *AQA Biology A-Level Year 2*. Coordination Group Publications (CGP), London.

Burrows, C., Lindle, C., McGarry, C., Pattison, S. Plowman, C., Rogers, R. & Thompson, H. (2015). *AQA Biology A-Level Year 1 and AS*. Coordination Group Publications (CGP), London.

A botanical digest.

Capon, B. (2010). *Botany for Gardeners*. Timber Press, London.

A simplified account of plant science.

Chalker-Scott, L. (2016). *How Plants Work*. Timber Press, London.

Written by a practising horticulturist.

Dawson, P. (1994). *A Handbook for Horticultural Students*. Peter Dawson, South Harting.

This is a detailed and extensive account of the wide range of topics embraced within the discipline of horticulture.

Dixon, G. R. & Aldous, D. E. (ed.) (2014). *Horticulture: Plants for People and Places: A Trilogy*, Volume 1: *Production Horticulture*, Volume 2: *Environmental Horticulture*, Volume 3: *Social Horticulture*, Springer Science + Business Media, Dordrecht.

A translation from German, providing detailed accounts of garden plant ecology.

Hansen, R. & Stahl, F. (1993). *Perennials and Their Garden Habitats*. (translated by Richard Ward). Cambridge University Press, Cambridge.

Written by a very knowledgeable practical gardener and still relevant.

Hellyer, A. G. L. (1952). *Your Garden Week by Week*. W. H. & L. Collingridge, London.

A useful and interesting account.

Hodge, G. (2013). *Botany for Gardeners: The Art and Science of Gardening Explained and Explored*. Octopus Publishing, London.

Several editions of this book explain the scientific principles underlying horticulture.

Ingram, D. S., Vince-Prue, D. & Gregory, P. J. (2015). *Science and the Garden: The Scientific Basis of Horticultural Practice*. John Wiley & Sons, London.

Still a readable account of science underpinning garden practice.

James, W. O. (1957). *The Background to Gardening*. George Allen & Unwin, London.

Although somewhat aged, this is still a very useable textbook.

Kimball, J. W. (1994). *Biology*. William Brown Publishers, Dubuque, Iowa.

Simplifies and de-mystifies botany.

Kratz, R. F. (ed.) (2011). *Botany for Dummies*. John Wiley & Sons, Chichester.

This author's books are authoritative.

Larkcom, J. (2002). *Grow Your Own Vegetables.* Francis Lincoln, London.
Larkcom, J. (2007). *Oriental Vegetables: The Complete Guide for Garden and Kitchen.* Francis Lincoln, London.
Larkcom, J. (2017). *The Salad Garden.* Francis Lincoln, London.

A long-standing textbook.

Lowson, J. M., Simon, E. W., Dormer, K. T. & Hartshorne, J. N. (1986). *Lowson's Textbook of Botany.* Unwin Hyman, London.

A recent digest that is readable.

Maumseth, J. D. (2016). *Botany: An Introduction to Plant Biology.* Jones & Bartlett Publishers, Bridgwater.

Explains the background to plants in the garden.

Niklas, K. (2016). *Plant Evolution: An Introduction to the History of Life.* University of Chicago Press, Chicago.

Now out of print, but a simple and straightforward text.

Norstog, K. & Meyerrieks, A. J. (1985). *Biology*, C. E. Merrill and A. Bell & Howell, Columbus.

A longstanding text, used in earlier editions.

Raven, P. H., Johnson, G. B., Mason, K. A., Losos, J. & Singer, S. (2016). *Biology.* WCB McGraw-Hill, Boston.

These books contains directions for a range of basic botanical experiments and is still relevant.

Thoday, D. (2015). *Botany: A Senior Text for Schools.* Cambridge University Press, Cambridge.
Wallis, C. J. (1966). *Practical Botany for Advanced Level and Intermediate Students: A Laboratory Manual.* William Heinemann, Medical Books, London.

Both Thoday and Yapp are readable textbooks. Yapp in particular has well stood the test of time.

Yapp, R. H. (2013). *Botany: A Junior Book for Schools.* Cambridge University Press, Cambridge.

Magazines and journals

Amateur Gardening, magazine published weekly and sold commercially, editor Gary Coward-Williams, Time Inc. UK, London.
Gardens Illustrated, magazine published monthly as a commercial magazine, editor Lucy Bellamy, Immediate Media Company, Bristol.
The Biologist, members' magazine of the Royal Society of Biology, published quarterly, editor Tom Ireland, London.
The Garden, The Royal Horticultural Society's members' magazine, published monthly, editor Chris Young, RHS Media, Peterborough.
The Plantsman, journal published quarterly and sold commercially, editor James Armitage, RHS Media, Peterborough.
Gardening websites
www.rhs.org.uk
www.greatbritishgardens.co.uk
www.ngs.org.uk (National Garden Scheme)

www.kew.org
www.bbc.co.uk/gardening
www.greatdixter.co.uk
www.jekkasherbfarm.co.uk
www.thegardeningwebsite.co.uk
www.bethchatto.co.uk
Biology websites
www.rsb.org.uk (Royal Society of Biology)
www.emergtoplifesci.org
http://botany.org/outreach/WebLinks.php
www.saps.org.uk (Science and Plants for Schools)
www.plantliffe.org.uk
http://bsbi.org (Botanical Society of Great Britain and Ireland)
www.kew.org (The Royal Botanic Garden at Kew is a source of both horticultural science and garden practice)
www.rbge.org.uk (The Royal Botanic Garden Edinburgh)
http://botanicgarden.wales (National Botanic Garden of Wales)
www.botanicgardens.ie (National Botanic Gardens of the Republic of Ireland)

All the main botanic gardens demonstrate expert garden practice and are centres for studies of horticultural science; local botanic gardens also provide examples of expert garden practice.

Apps: an increasingly helpful and expanding source of gardening and scientific information.
Growit!
Gardenanswersplantidentifier
SmartPlant
GardenTags

Seed, fruit and bulb catalogue websites (this is a sample of companies, there are many other companies that serve gardens and gardening; it would not be feasible list more, and omissions are not to be construed as comments on quality or customer service).

www.thompson-morgan.com
www.mr-fothergills.co.uk
www.marshalls-seeds.co.uk
www.PomonaFruits.co.uk
www.dutchbulbs.co.uk
www.davidaustinroses.co.uk
www.claireaustin-hardyplants.co.uk (perennial plants)
www.suttons.co.uk
www.unwins.co.uk
www.kingsseeds.com
www.kenmuir.co.uk (soft and tree fruit)
www.frankpmatthews.com (fruit trees)
www.eurobulbs.co.uk

Index

For Product Safety Concerns and Information please contact our EU representative GPSR@taylorandfrancis.com Taylor & Francis Verlag GmbH, Kaufingerstraße 24, 80331 München, Germany

Batch number: 10777132

Printed by Printforce, the Netherlands